The Chemical and Pharmaceutical Industry in China

G. Festel · A. Kreimeyer
U. Oels · M. v. Zedtwitz

Editors

The Chemical
and Pharmaceutical Industry
in China

Opportunities and Threats
for Foreign Companies

With 29 Figures
and 81 Tables

 Springer

Dr. Gunter Festel
Schuermattstrasse 1
6331 Huenenberg
Switzerland
E-mail: gunter.festel@festel.com

Dr. Andreas Kreimeyer
Vorstand BASF Aktiengesellschaft
67056 Ludwigshafen
Germany
E-mail: andreas.kreimeyer@basf-ag.de

Dr. Udo Oels
Vorstand Bayer AG
Gebäude W 11
51368 Leverkusen
Germany
E-mail: udo.oels.uo@bayer-ag.de

Professor Dr. Max von Zedtwitz
Mailbox B-55
School of Economics & Management
Tsinghua University
Beijing 100084
P.R. China
E-mail: max@post.harvard.edu

Cataloging-in-Publication Data

ISBN 978-3-642-06140-0   e-ISBN 978-3-540-26561-0

Springer is a part of Springer Science+Business Media
springeronline.com

© Springer-Verlag Berlin Heidelberg 2010
Printed in Germany

Cover design: design & production GmbH

Preface

"Don't miss out on China!" and "What are you doing about China?" Catch phrases like these are spreading among managers all over the world. Just take a brief look at the business class occupancy of flights from Europe, North America or Japan to major Chinese cities: This gives you a glimpse of how business people are attracted by steady growth rates of 6 percent to 10 percent. It also indicates how much attention is given to a market featuring 1.3 billion potential consumers and a government committed to rapidly changing the country from an agriculture-dominated developing country into one of the world's economic powerhouses. Most of the global industrial players have had economic ties with China for decades already, but they were further strengthened after the country's opening to the world in the early 1980s. Furthermore, China's accession to the World Trade Organization is expected to catapult this already surging economy into another sphere of development.

The Chinese market is of increasing significance to the global chemical and pharmaceutical industry. This book analyzes and illustrates the current situation from different viewpoints. It is structured in three parts: the first two parts focus on a characterization of the chemical and pharmaceutical industries in China, outlining the economic and political situation in China with its strategic and operational implications for Western companies, complemented with strategy-in-action cases of BASF, Degussa, Merck and Novartis. The third part of the book comprises case studies describing how Western companies like BASF, Bayer, Bicoll, Ciba, Degussa, and DSM are managing the market entry and investments for further growth in China. These are more than mere testimonials: these case studies are rich compendiums of good and best practices of launching new businesses in China. Based on thorough analysis the authors disclose and provide new strategies, approaches and tools to deal with sometimes not obvious challenges in China. The book also analyzes the threat to Western companies in their home markets from Chinese competitors.

We are honored to have a long list of distinguished experts contributing as authors to this book. We would like to thank the authors for their time, effort and commitment in formulating their insights and putting them down in writing. We are also grateful to Gerhard Sichelstiel, Martin Jung, Joachim Luithle and Elaine Britton, whose invaluable contributions to this unique compilation have made this book a reality. Last but not least, we would like to thank Martina Bihn of Springer Verlag for her support and editorial guidance.

Gunter Festel   Andreas Kreimeyer   Udo Oels   Maximilian von Zedtwitz
Huenenberg   Ludwigshafen   Leverkusen   Beijing

March 2005

Preface

"Don't miss out on China!" and "What are you doing about China?" Catch phrases like these are spreading among managers all over the world. Just take a brief look at the business class occupancy of flights from Europe, North America or Japan to major Chinese cities: This gives you a glimpse of how business people are attracted by steady growth rates of 6 percent to 10 percent. It also indicates how much attention is given to a market featuring 1.3 billion potential consumers and a government committed to rapidly changing the country from an agriculture-dominated developing country into one of the world's economic powerhouses. Most of the global industrial players have had economic ties with China for decades already, but they were further strengthened after the country's opening to the world in the early 1980s. Furthermore, China's accession to the World Trade Organization is expected to catapult this already surging economy into another sphere of development.

The Chinese market is of increasing significance to the global chemical and pharmaceutical industry. This book analyzes and illustrates the current situation from different viewpoints. It is structured in three parts: the first two parts focus on a characterization of the chemical and pharmaceutical industries in China, outlining the economic and political situation in China with its strategic and operational implications for Western companies, complemented with strategy-in-action cases of BASF, Degussa, Merck and Novartis. The third part of the book comprises case studies describing how Western companies like BASF, Bayer, Bicoll, Ciba, Degussa, and DSM are managing the market entry and investments for further growth in China. These are more than mere testimonials: these case studies are rich compendiums of good and best practices of launching new businesses in China. Based on thorough analysis the authors disclose and provide new strategies, approaches and tools to deal with sometimes not obvious challenges in China. The book also analyzes the threat to Western companies in their home markets from Chinese competitors.

We are honored to have a long list of distinguished experts contributing as authors to this book. We would like to thank the authors for their time, effort and commitment in formulating their insights and putting them down in writing. We are also grateful to Gerhard Sichelstiel, Martin Jung, Joachim Luithle and Elaine Britton, whose invaluable contributions to this unique compilation have made this book a reality. Last but not least, we would like to thank Martina Bihn of Springer Verlag for her support and editorial guidance.

Gunter Festel	Andreas Kreimeyer	Udo Oels	Maximilian von Zedtwitz
Huenenberg	Ludwigshafen	Leverkusen	Beijing

March 2005

Table of Contents

Part I: The Chemical Industry in China

1 The Global Chemical Industry

Gunter Festel: Festel Capital, Huenenberg, Switzerland

This chapter describes the global chemical industry as well as major trends like the ongoing transformation of the industry and industrial biotechnology which will have a big impact, especially in emerging economies like China.

1.1 Some Basic Facts About the Chemical Industry

The chemical industry today is one of the largest and most diversified industries in the world with an impressive history (*Arora/Landau/Rosenberg* 1998). More than 1,000 large and medium-sized companies manufacture over 70,000 different product lines, and the total value of chemicals produced in 2002 was EUR 1,841 billion (including pharmaceuticals). The chemical industry accounts for 3 to 4 percent of global gross domestic product (GDP), which probably makes it the largest manufacturing industry in the world. The European Union is the largest market with 29 percent, followed by the United States with 26 percent, Japan with 10 percent and China with 6 percent (*Fig. 1.1*). Within the European Union, Germany is the largest national market with sales of EUR 133 billion in 2002 and EUR 136 billion in 2003.

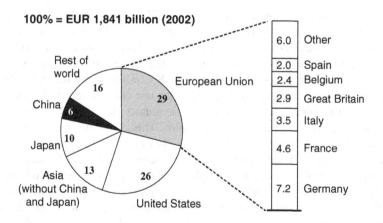

Fig. 1.1. Global chemical market (including pharmaceuticals) by region, 2002 (percent) (Source: CEFIC)

In emerging economies, and particularly in those of China and South Korea, the consumption of chemical products accounts for about 10 percent of GDP and there is significant potential for further growth in these countries. In mature economies such as Germany and the United States, however, future growth rates of chemical consumption are falling below those of GDP.

Chemical companies serve almost every single industrial sector – from food processing to construction and electronics. The automotive industry, for example, relies on many chemical products in the production of tires, seats, dashboards and coatings. Industries like electronics and pharmaceuticals are also important customers in which innovation is often driven by chemicals such as new materials for electronic devices or new active substances for pharmaceuticals.

The chemical industry has a distinctly multiregional character in the sense that there are limited trade flows between the three main manufacturing regions of North America, Europe and Asia. Only about 10 percent of total output is shipped between these three regions. Not surprisingly, interregional trade is particularly limited for volume products which are relatively expensive to transport. The limited interregional trade has not, however, decoupled prices and industry cycles in the different regions. The prices of commodity plastics, for example, moved in remarkably close harmony in all three regions. Unlike interregional trade, trade within the regions is very strong. In Europe, for instance, 46 percent of total chemical industry output in 2002 was exported within the European Union itself. 29 percent was exported outside the European Union and only 25 percent was consumed in the country of manufacture.

1.2 Transformation of the Chemical Industry

The chemical industry consists of three groups of companies (*Fig. 1.2*). The first group is focused on commodity chemicals such as basic chemicals or plastics and accounts for a little over one third of total sales. These companies are often the chemical subsidiaries of the big oil companies like BP, ExxonMobil and Shell.

The second group of companies like Ciba Specialties, Clariant and Rohm & Haas is focused on specialties and accounts for just one quarter of all sales. The third group, the integrated conglomerates, dominates the chemical industry and contributes almost 40 percent of overall sales. This conglomerate structure was characteristic for the chemical industry. Nearly every industrialized country has the residue of its own nationally developed chemicals industry in which every company traditionally tried to produce the whole range of chemical products. Especially during the 1970s and 1980s these companies grew to considerable size and still have a large number of different businesses, ranging from upstream basic chemicals, specialty chemicals, agrochemicals and in some cases also pharmaceuticals (examples are Akzo Nobel and Bayer).

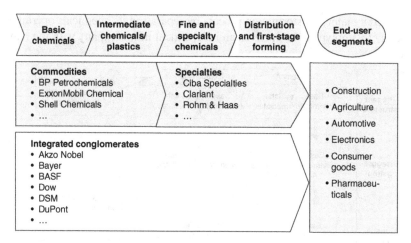

Fig. 1.2. Structure of the chemical industry today

Due to their large and diversified portfolios, it is increasingly difficult for the conglomerates to provide all the resources necessary to run them. Synergies between chemical and other businesses seem to be either non-existent or difficult to capture so that specialized players generally perform better than non-specialized players. Therefore, investors are more interested in focused companies with less complicated and more transparent structures. But the conventional wisdom used to be that specialties are better than commodities, and that less cyclical businesses are better than more cyclical ones is in the longer term not true. The top quartile of commodity firms and specialty firms outperform by far all categories in the lower quartiles. The key factor, it seems, is operational excellence, regardless of the nature of the business. That means that excellent commodity firms are better value creators than average specialty players.

Many of the diversified conglomerates began radical changes during the 1990s, kicking off a global industry restructuring and M&A wave in an attempt to exit and divest those businesses without leading positions (*Fig. 1.3*). The complex conglomerates containing chemical and non-chemical businesses under one roof started to break up by spinning-off non-core businesses and focusing their portfolios on attractive product segments (*Festel* 2003a, 2003b). In that sense a lot of major transactions have taken place in the last ten years. BASF sold its chemical trading activities, magnetic tapes business and pharmaceuticals operations (Knoll). DOW also divested its pharmaceuticals business. DSM sold its petrochemical business and strengthened its life science business by buying Gist Brocade's and Roche's vitamins divisions. The most spectacular transaction was the break-up of Hoechst: after divesting its specialty chemicals business to Clariant, Hoechst spun-off its basic chemicals businesses (now Celanese) and merged its life science divisions (pharmaceuticals, animal health, crop science) with Rhône-Poulenc to found the new company Aventis. Aventis (which was recently bought by Sanofi) has since sold its crop science business to Bayer which last year started a major transformation process by separating its chemicals operations and parts of its polymers business as an independent new company with the name Lanxess.

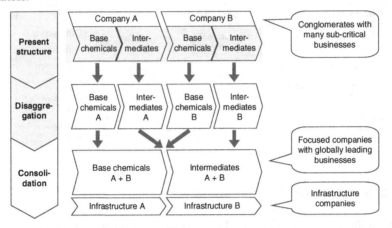

Fig. 1.3. Trend from conglomerates with many sub-critical businesses to focused companies with globally leading businesses

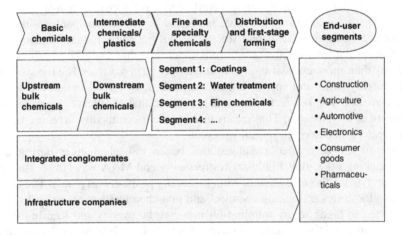

Fig. 1.4. Structure of the chemical industry tomorrow

At the end of this transformation process, there will be more pure players producing bulk chemicals in the upstream (e.g. ExxonMobil Chemical, Shell Chemicals) or downstream (e.g. Basell, Borealis) commodity value chain. On the specialties side, focused players will start to form around segments like coatings, water treatment or fine chemicals. Infrastructure is no longer the core business of these focused companies and was divested during recent years so that a new class of independent infrastructure companies (e.g. Infraserv Höchst) appeared (*Fig. 1.4*).

1.3 Impact of Industrial Biotechnology

Scientific breakthroughs in the life sciences are fascinating and will change the structure of the industry (*Enriquez/Goldberg* 2000). Biotechnology has already demonstrated that it can accelerate innovation in pharmaceuticals, where its application is most advanced. Over the past ten years, most of the truly innovative drugs (those that address an unmet medical need) have come from the application of biotechnology. Many more biotech-based products are currently in the pipeline and are expected on the market soon. But what are the implications of biotechnology for the chemical industry and how will they change the industrial landscape?

The market for biotech-based products, excluding ethanol and starch derivatives which have traditionally been produced by fermentation, accounts today for only 3 percent of the total chemical market (about US$ 30 billion). Examples include biocatalysis and biomolecules in fine chemicals, biopolymers as substitutes for synthetic polymers, enzymes and modified additives in specialties, and the production of basic and intermediate organic chemicals using fermentation. The share of biotech-based chemicals will increase dramatically within the next years. In 5 to 10 years (2010+) nearly 20 percent of chemical products with a sales volume of approximately US$ 300 billion will be produced by biotechnological processes (*Fig. 1.5*).

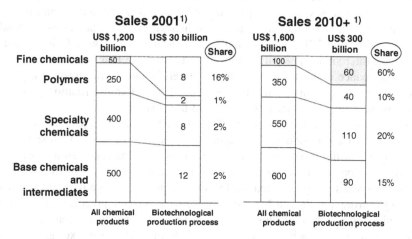

Fig. 1.5. Importance of biotechnological production processes in 2001 and 2010+

Fine chemicals offer the greatest potential (60 percent) using biotech routes for the production of enantiomerically pure complex chiral molecules (*Festel/Knöll/Götz/Zinke* 2004). More than 50 percent of the top 100 drugs are based upon enantiomerically pure molecules and such drugs today already post sales exceeding

US$ 100 billion. In addition, 60 percent of the new active pharmaceutical ingredients in drug development phases II and III are chiral and 90 percent of the new chiral substances are enantiomerically pure. Also, specialty chemicals (20 percent) will be increasingly produced using biotech processes with many potential applications in the food, cosmetics, textile and leather industries. Within polymers (share of 10 percent) as well as basic chemicals and intermediates (share of 15 percent), a key focus will be on using renewable raw materials and new products which are not available by chemical synthesis.

The most important biotechnological production processes in the coming years will be the fermentation of microorganisms and biotransformation. Fermentation with animal/plant cell cultures is especially suitable for the synthesis of complex biomolecules such as sugar- and lipid-modified molecules and proteins which are not accessible by chemical synthesis routes. Although fermentation requires high dilution, it is a single step process that leads to the desired product.

In contrast, biotransformation can work at concentrations up to 50 percent but a single enzymatic step does not necessarily lead to the desired product, so the reaction may need to be performed in multiple steps. However, technical development has progressed rapidly and will further boost the application potential of biotransformation. Increasing knowledge of enzyme reactions in non-aqueous solutions will lead to a broader spectrum of processes and a greater number of potential substrates. Process efficiency will be improved due to new developments in reactor and process design. The versatile application of extremophiles will enable more robust processes, increasing the diversity of process conditions and reducing reaction times. Additionally, the use of directed evolution processes can lead to the development of tailor-made and high-performance enzymes. Genetic engineering of microorganisms will result in the discovery of new enzymes and reactions and reduce unwanted side reactions in cell-bound biotransformation processes. As a result of this technical progress cell-free biotransformation (enzyme catalysis) will widely replace fermentation in the next ten years.

Restrictions in the use of biotechnological production processes are primarily seen on the economic side, e.g. operating costs, R&D costs and investment. The synthesis of existing products by chemical processes is frequently so inexpensive that the development of a biotechnological production process is often not cost–efficient, especially if production facilities for chemical routes already exist. This is a big hurdle for the further establishment of biotech processes in mature economies. However, this hurdle is much lower in emerging economies like China where new capacities have to be built to meet strong market growth. In these cases, biotech processes could provide clear cost advantages as they often have substantially lower capital and manufacturing costs and allow greater flexibility due to lower minimum economies of scale. Additionally, biotechnology is more eco-friendly with less waste produced and less energy consumed. Therefore, biotech will have a bright future in chemicals production – especially in China.

2 The Petrochemical Industry in China

Jörg Wuttke: BASF China Ltd., Beijing, P.R. China

2.1 Major Players in the Chinese Chemical Industry

The development of China's petroleum and petrochemical industry can be roughly phased into three stages. The first stage started with the discovery of the famous Daqing Oil Field in northern Heilongjiang province and ended in 1978. During this stage several large-scale oil fields were developed and the capacities of some oil refining enterprises in the provinces of Gansu, Liaoning and Shandong were expanded. Yanshan Petrochemical Company in Beijing became China's first enterprise to boast ethylene production facilities with an annual capacity of 300,000 tons. This was the starting point of the petrochemical industry in China.

The second stage from 1978 to 1998 saw the establishment of a series of large-scale petrochemical industrial bases in the country. They are in Daqing, Yangzi (Jiangsu), Qilu (Shandong), Shanghai, Jilin and Maoming (Guangdong).

The period since 1998 is viewed as the third stage. In that year the government divided the assets. The northern region with the larger oil reserves went to China National Petroleum Corp. (CNPC), whereas the southern region with less resources yet the large coastal market went to China Petroleum and Chemical Industry Corp. (Sinopec) (see *Fig. 2.1*). China National Offshore and Oil Corp. (CNOOC) was kept offshore. The three local giants in the industry, namely CNPC, Sinopec and CNOOC were successfully listed on the stock exchanges in London, New York and Hong Kong in 2000, and their listed companies are called PetroChina, Sinopec Corp. and CNOOC Ltd. respectively.

Two of the three corporations (CNPC and Sinopec) have formed a complete business system ranging from oil and gas exploitation to sales of finished products. Sinopec and CNPC dominate production of ethylene, petrochemical intermediates and downstream products in China, with their combined capacity accounting for over 90 percent of the country's total. Of the 18 ethylene plants in China, 16 are owned by the two giants. The future presence of CNOOC in the sector with the joint construction with Shell of an ethylene plant producing 800,000 tons a year at Nanhai, Guangdong province will alter the current pattern of capacity distribution.

Local medium-sized (e.g. Huayi Corporation in Shanghai) and small state enterprises and a large number of township and private enterprises are mainly active in downstream petrochemical segments, while foreign investment focuses on plastics and synthetic fibers that are in short supply in China and highly dependent on imported raw materials. For domestic synthetic resin production, Sinopec and CNPC contribute the vast majority of polyethylene (PE), polypropylene (PP) and synthetic rubber, but only 10 percent of polyvinyl chloride (PVC) for which production quotas were historically allocated to local state chemical enterprises (e.g. Shide in Dalian, Liaoning Province) and dependent on imported raw materials from various sources. Sinopec and CNPC supply the majority of synthetic fiber intermediates such as purified terephthalic acid (PTA), ethylene glycol (EG), acrylonitrile (AN) and caprolactam (CPL), while local state companies and non-state enterprises produce synthetic fibers including polyester, acrylic, polyamide and polypropylene fibers. Foreign joint ventures (e.g. BASF, LG Chemical, ChiMei) manufacture most of the country's polystyrene (PS) and ABS.

Fig. 2.1. Geographical division of business operations of CNPC and Sinopec (Source: CNPC and Sinopec)

2.2 Underpinning Economic Growth and the Energy Bottleneck

The Chinese economy grew strongly following the start of reforms in 1978 and the opening in the early 1980s. Although there have been several ups and downs, the average growth rate of more than 8 percent is unprecedented for an economy of this size and complexity (source: China's State Statistical Bureau). The outlook of 6 to 7 percent growth in GDP until 2015 is based on models provided by BASF (*Fig. 2.2*).

GDP growth in %

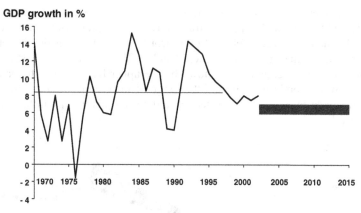

Fig. 2.2. Continued robust GDP growth of China (Source: World Bank, BASF)

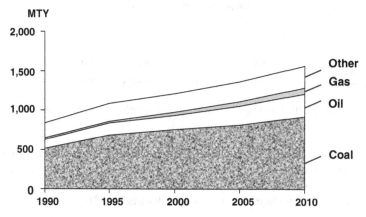

Fig. 2.3. In China's fuel mix, coal remains dominant but oil is growing (Source: ExxonMobil Data 2000, Stephen F. Goldmann)

This rapid economic growth has led and will lead to an increasing demand for energy. The outlook for China through 2010 broken down by fuel type including "other", (i.e. primarily hydroelectric and nuclear) is based on many assumptions to project the fuel mix. In particular, the pace of natural gas development and the mix of fuels for power generation are difficult to predict (*Fig. 2.3*). Coal will remain the dominant fuel throughout the period. However, the energy mix will change because the growth of oil and gas is greater than that of coal. Oil demand is growing at a rate of 4 to 5 percent compared with 2 percent for coal during this decade. This leads to an analysis of the oil supply/demand, which is important for the feedstock analysis for the chemical industry (*Fig. 2.4*).

There will be a growing dependency on imports. The lower line on the chart shows the outlook for domestic Chinese oil production. It is relatively flat across the period. The shaded area shows the likely range of uncertainty. It is relatively narrow since, given the long development lead times, any substantial volumes that could be brought to market in this time frame have mostly been discovered and are under development. Even this relatively flat profile represents considerable effort in offsetting the decline in mature domestic fields. There is more uncertainty on the demand side of the equation, as shown by the upper lines. This is fundamentally driven by economic growth, and the range of uncertainty reflects differing assumptions made by various forecasters regarding the drivers that influence the demand for oil.

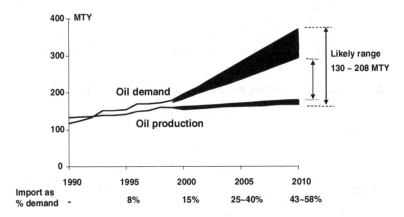

Fig. 2.4. Growing oil imports likely (Source: ExxonMobil Data 2000, Stephen F. Goldmann)

China's import dependency will continue to grow from around 40 percent today (2003: 91 million tons (MT) of oil, 1.8 million barrels a day (MBD)) to between 43 and 58 percent in 2010.

According to various forecasts, China's oil demand in 2010 is estimated to be:

- 300 MT or 6 MBD by Chinese government assessment
- 346 MT or 7 MBD by the IEA (International Energy Agency)
- 358 MT or 7.2 MBD by CERA (Source: CERA, Cambridge Energy Research Associates 2002)

This translates into imports of between 130 million tons per year (MTY) (2.6 MBD, import dependency 43 percent) according to the Chinese government forecast, or 208 MTY (4.2 MBD, import dependency 58 percent) according to CERA.

China's economic growth has outstripped its own energy resources. When it comes to consumption China is an energy superpower second only to the United States. Meeting its energy needs is one of the most difficult challenges China will face. Anything above 130 MTY (2.6 MBD) is a large number. Nevertheless, compared with Western Europe's oil demand of about 700 MTY in 2002 and imports of 400 MTY it might not seem particularly large in absolute numbers. However, the market would be concerned about that *additional* amount in a tight oil market but can accommodate it in terms of availability. It is a matter of setting up new commercial relationships without conflict. The International Energy Agency (IEA) believes China will become a far more influential player in the global oil market, specifically in the oil-rich Middle East. A most significant result of this in terms of facility planning is that the average crude slate will become heavier and higher in sulfur content, that is, more sour.

What is the outlook with respect to the supply and demand of petroleum products and the implications in terms of refining capacity and the subsequent feedstock situation for the chemical industry?

2.3 Outlook for Feedstock Supply

Fig. 2.5 shows refinery supply and demand, with the tops of the columns representing total product demand consistent with the outlook discussed earlier.

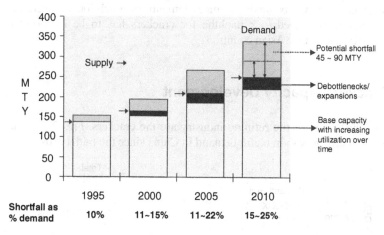

Fig. 2.5. Growing petroleum product shortfall (Source: ExxonMobil Data 2000, Stephen F. Goldmann)

It can be seen that demand is met from three components. The first is the existing capacity, which increases over time with the assumption that capacity utilization will increase from the present level of 70 percent to about 90 percent by 2010, consistent with best practices in other parts of the world. There will also be some debottlenecking and expansions. The balance will be met by product imports, which would be in excess of 45 MTY or 1 MBD depending on how much demand increases. The need for additional refining capacity to process increasing volumes of heavier, high sulfur crude will require significant investments and lead times for facility planning. The average scale of China's refineries is much lower than the world average. Yet there are notable exceptions such as the Sinopec Zhenhai Refinery with a capacity of 16 MTY. The structure of refining equipment is biased towards fluid catalytic crackers, which account for 35 percent of total refining equipment as compared to the global average of 17 percent. It is also biased towards oil with a low sulfur content, so the capacity for processing sulfur oil is insufficient.

Within China it is mainly the east and southeast that will continue to be short of refining capacity while the northeast and northwest will remain product exporters to the high-demand sectors, even after allowing for some planned shutdown of small and inefficient capacities. The greatest requirement for capacity additions is in the rapidly growing provinces of the south and southeast. Overall, China will have a refining capacity shortfall by 2010 even after allowing for a dramatic improvement in capacity utilization and some expansions and debottlenecking. China's naphtha demand is expected to surge by 67 percent by 2005 to 34.5 million tons, spurred by strong demand from the petrochemical manufacturing sector, especially along the coastline. And demand is likely to rise further to 53 million tons by 2010 and 90 million tons by 2020. China currently produces about 21 million tons of naphtha a year, which basically meets local demand. But with production capacities for ethylene (a naphtha product) set to increase, domestic naphtha production will be unable to meet demand for much longer, according to official sources. The feedstock naphtha for crackers has to be imported from Korea, Taiwan and other Asian countries.

2.4 Cracker Capacity Development

The main customers of the refining industry are the crackers. *Fig. 2.6* illustrates the very dramatic growth in olefin demand in China since the mid-1980s.

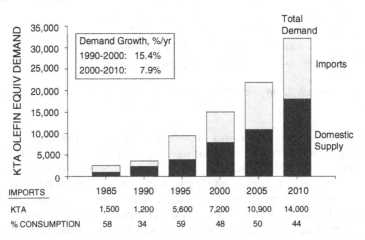

Fig. 2.6. Growing olefin derivatives imports in China (Source: ExxonMobil Data 2000, Stephen F. Goldmann)

This is very consistent with the growth observed in other rapidly developing economies around the world. Demand in 2010 is projected to be about ten times the demand in 1985. In spite of very rapid growth in domestic supplies, China will remain about 50 percent import-dependent for olefins in 2010. This outlook assumes that all of the six major grassroots steam cracker projects currently under construction and discussion will be in operation by 2010. These are BASF Yangzi Corporation in Nanjing, BP Sinopec Caojing (SECCO), Shell CNOOC Nanhai, and possible domestic crackers operated by Sinopec in Tianjin, Zhenhai and Caojing No.2.

Even with the planned debottlenecking of existing smaller crackers and possibly six new grassroots additions, China remains largely import-dependent (see *Fig. 2.7*). This conclusion remains even if more conservative assumptions are made about GDP growth. Even if GDP increases at the unlikely low figure of only 5 percent annually for the next decade, China will still need to import about 30 percent of its ethylene demand. There are major crackers coming on stream in the Middle East over the next three years, and China is their main target market. If China eventually tries to achieve self-sufficiency, this could cause world market prices to erode.

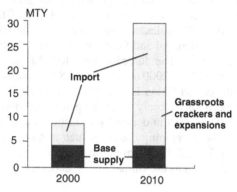

Fig. 2.7. Possible scenario of ethylene capacity development (2010 vs. 2000) (Source: China Petroleum & Chemical Industry Association (CPCIA), BASF)

2.5 Consumption of Ethylene and Its Derivatives

From 1995 to 2003, China's demand for ethylene increased at an average annual rate of more than 12.5 percent. China's national ethylene production capacity reached 6.12 million tons in 2003, but domestic demand climbed to 15.4 million tons. As locally made ethylene and downstream products are in short supply, petrochemical imports to China have increased by an average 32 percent since the 1990s. Foreign raw materials and products have accounted for more than 50 percent of domestic demand for synthetic resins and of total consumption of ethylene equivalent petrochemicals. The figure exceeds 80 percent for polystyrene and ABS consumption (Source China's State Statistical Bureau (Beijing)).

Another feature of ethylene and downstream products is that their production and consumption is growing faster than GDP. The average GDP growth rate in

China for the period 1990 to 2003 was about 9 percent while the consumption of the five main synthetic resins grew from 2.27 million tons to 16.60 million tons, yielding average annual growth of 22 percent. The demand for synthetic rubber leaped from 320,000 tons to 940,000 tons at an average rate of 13 percent. Synthetic rubber consumption increased from 1.65 million tons to 6.62 million tons, with an annual growth rate of 17 percent.

2.5.1 Forecast of Market Demand for Ethylene in China

Due to the comparatively mature marketing mechanisms in the Western countries, it is easier to make predictions in the United States or Europe. Companies there only have to consider four factors: historical inertia, ultimate purpose, new applications and substitute products. The elastic coefficient is usually set at between 1 and 1.2. In China, the market for petrochemical products is still growing rapidly and less predictably. The country's ethylene consumption grew at an annual rate of 12.5 percent during the period from 1995 to 2003, while its GDP increased at a rate of less than 10 percent. The calculated annual elastic efficiency ratio between the growth rate of ethylene demand and GDP for the period of 1995 to 1997 was 2.1, 2.4 and 1.9 respectively. The forecast predicted that this efficiency ratio should be 1.0 for the period from 2000 to 2005 and 0.8 for 2005 to 2010.

The demand for ethylene in China is mainly driven by high demand for plastics. This has always been very closely linked to general economic activity. As a general rule of thumb, the growth in demand for plastics is a multiple of the GDP growth of the country. For developing countries such as China, demand elasticity has often been around 1.5 to 2.5 times that of the country's real GDP growth while in a more mature country such as the United States, elasticity is often lower at 0.6 to 0.8 times.

Therefore, 60 percent of the ethylene produced in China is used to make polyethylene, and the proportion is slightly higher than the world average of 56 percent. 13 percent are used to manufacture glycol, 10 percent for PVC, 3 percent for glacial acetic acid and 18 percent for other products. Actual ethylene demand in China in 2003 was 15.4 million tons. The country is likely to need 18.5 million tons of ethylene in 2005 and 28.5 million tons in 2010.

2.5.2 Forecast of Market Demand for Ethylene Derivatives in China

The demand for ethylene translates to a development in demand for three main categories of synthetic materials: synthetic resins, synthetic rubber and synthetic fibers.

Synthetic Resins

Polyethylene, polypropylene, polystyrene, PVC and ABS the most widely used, and their consumption accounts for over 80 percent of total synthetic resin consumption in China. Polymethylmethacrylate (PMMA) and polyformaldehyde (POM) are used to a much lesser extent. To predict the market for plastic products in China, the following development trends are important.

- Farm-use plastic products still enjoy a vast market but the growth rate is declining.
- Demand for resins for construction applications will climb quickly.
- Demand for resin for packaging applications will continue to grow at a high rate.
- Demand for utensil resins will decline.
- Demand for synthetic materials for electronic products, wiring and household appliances will see higher growth.
- Plastics used in automotive applications will experience stable market growth.
- Demand for resin materials for the production of disposable meal boxes and trays will see declining growth rates.
- Demand for resins for larger household appliances will grow at a slower rate.

Synthetic Rubber

Synthetic rubber products mainly include styrene butadiene rubber (SBR), cis-1,4-polybutadiene rubber (BR), neoprene (CR), Buna-N rubber (NBR), ethylene propylene rubber (EPR), butyl rubber and isoprene. The first two types of rubber account for 83 percent of the country's total synthetic rubber production capacity.

China has very limited natural rubber resources, with output stabilized at 500,000 to 600,000 tons a year. However, the country's annual consumption of such materials is 1.5 to 1.6 million tons. In 2003, China produced 188 million tires, of which approximately 25 percent were exported.

In predicting the market for synthetic rubber, the following forecast can be made:

- Major consumption comes from the tire production sector.
- Radial ply tires are taking a larger share in total tire production, which will cut demand for synthetic rubber.
- Half of the existing rubber-processing enterprises are Sino-overseas joint ventures which require rubber of higher quality. Most of these producers rely on imported rubber materials.

Synthetic Fibers

China is currently the world's largest producer and consumer of polyester fibers (for textiles). According to government statistics, China's 2003 consumption of purified terephthalic acid (PTA) - the primary raw material used in the manufacture of polyester fibers for textiles - and polyethylene terephthalate (PET) for bottles, textiles, packaging and film products exceeded 8 million tons. However, of that amount, only 30 percent was supplied from domestic production. Fibers are mainly used for civilian purposes with little volume going to industrial sectors. Synthetic fibers include polyester, acrylic, polypropylene, polyamide and vinylon fibers. Over 60 percent of polyester fibers are used in the production of outerwear, 20 percent for decorative fabrics and 15 percent for industrial textiles. Acrylic fibers are used in the production of wool blend yarns and carpets. Polyamide fibers are used to make flat chafer fabrics and lining fabrics for clothes. Polypropylene and vinylon are used in industrial applications. China's fiber-processing volume has increased in the last few years and reached 13.7 million tons in 2003, the largest in the world. The current production of synthetic fibers in China has reached 11.8 million tons, making the country number one in the world.

The market for synthetic fibers is following the development trends described below:

- Most of the synthetic fibers are used for clothing and the change in fashion as well as purchasing power will determine the growth in demand for such fibers. As the standard of living improves, demand for personalized and functional fibers for clothing will grow at the cost of low-quality products.
- Due to the limited of natural fibers, synthetic fibers will play the leading role. In China, natural and synthetic fibers share the market equally. However, the output of natural fibers has stabilized at five million tons and will not increase noticeably in the coming years.
- Acrylic fiber products are now at a mature stage of the market cycle and demand for these will slow. New applications for polyester and polyamide fibers will post growth in the coming years.

Table 2.1 shows the predictions made by experts for the demand for ethylene and related products in the coming ten years.

Table 2.1. Predicted demand for ethylene and related products (million tons) (Source: BASF)

Product	2005	2010	2015
Ethylene	18.5	28.5	>30
Five categories of synthetic resins	21.6	28.6	37
Synthetic rubbers	1.1	1.4	1.7
Synthetic fibers	8.2	9.8	11

2.6 WTO Entry and Its Impact on China's Petrochemical Industry

The situation in China since entering the WTO is that with less policy protection from state authorities, multinational companies are going to compete with their local counterparts on the basis of their significant advantages in funding, technological R&D, human resources, service, distribution skills and brands. How soon Chinese petrochemical producers can enhance their competitiveness will be vital for them to survive the challenges. China's duty on crude oil imports has been eliminated and the duty tariffs on the following downstream products are listed in *Table 2.2*.

Table 2.2. Import tariff development according to China's WTO 2001 accession commitment (Source: US-PRC WTO Agreement, websites)

	2000 tariff (%)	WTO agreed terms	
		Tariff (%)	Effective year
Crude oil	RMB16 per ton	0	2000
Naphtha	6	6	2000
PE	18	6.5	2008
PP	16	6.5	2008
PS	16	6.5	2008
Styrene	9	2	2005
ABS	16	6.5	2008
Ethylene	5	2	2003
Ethylene glycol	14	7	2003
Acrylic esters	9	6.5	2001
Acrylic acid	9	6.5	2001
Methyl amines	9	6.5	2001
C4-oxo alcohols	8	5.5	2001
Formic acid	9	6.5	2001
Propionic acid	6	5.5	2000
DMF	8	6.5	2001
Isocyanates	10	6.5	2003
Nylon-66	16	6.5	2005
Nylon-6	10	9	2005
Polyester	19	5	2005

2.7 Competitive Environment in the Chinese Market

China's market for petrochemical and chemical products is served by the large petrochemical producers (PetroChina, Sinopec, CNOOC), by imports brought in by many local fragmented companies and by some multinationals. The individual market shares differ at the individual levels of the value chains (see *Fig. 2.8*).

The Chinese petrochemical and chemical industry is still dominated by Sinopec (chemical business worth RMB 92 billion in 2003), PetroChina (chemical business worth RMB 39 billion in 2003) and CNOOC (at this stage virtually no chemical business). They are the big three in China. Yet other companies are getting stronger either by means of M&A (e.g. ChinaChem) or organic growth. These Chinese companies might become very strong competitors in their respective fields (see *Table 2.3*).

Roughly one third of China's oil products are imported. Around 75 percent of imports come from neighboring counties like Singapore and the Republic of Korea (ROK). For example, the ROK, Japan and Taiwan Province together account for 66 percent of general-purpose resin imports to China every year. Japan, ROK, Singapore, Malaysia and Taiwan Province also contribute 80 percent of China's polyester imports. About 90 percent of China's styrene butadiene rubber imports came from Japan, ROK, Russia and Taiwan Province.

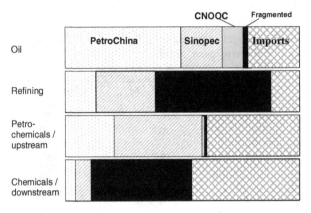

Fig. 2.8. Structure of the petrochemical and chemical market in China (Source: BASF estimate based on different analysts' reports and Chinese statistics)

Table 2.3. Top 10 chemical companies in China (Source: China Petroleum and Chemical Industry Association (CPCIA); 2003 annual reports of PetroChina, Sinopec Corp. and CNOOC Co. Ltd.)

Rank	Company	Turnover in 2003 (in RMB million)
1	PetroChina (listed part of CNPC)	455,133
2	Sinopec Corp. (listed part of Sinopec Group)	443,136
3	CNOOC Co. Ltd. (listed part of CNOOC Group)	40,950
4	Shandong Binghua Group Co. Ltd.	7,795
5	China National Chemical Industry Corp.** (ChemChina)	6,900
6	Liaoning Huajing Chemical (Group) Co. Ltd.	3,505
7	Jiangsu Lingguang Group	3,503
8	Juhua Chemical Group	2,850
9	Jiangsu Chemicals and Pesticides Group Company	2,035
10	Yantai Wanhua Polyurethanes Co. Ltd.	2,008

* Rubber processors, chemical fertilizer and chloro-alkali producers are excluded from the list.
** Annual sales based on Bluestar + Haohua

Most European and American companies compete for China's market by establishing joint ventures. The local players will have to cope with higher pressure mainly from overseas companies in the foreseeable future. This is difficult as domestic facilities are mostly sub-scale, e.g. of 18 ethylene plants, only seven produce more than 300,000 tons per year, compared with the world average of 750,000 tons per year.

During the coming years the competitive environment in China will change. The country's petrochemical industry is currently in a phase of restructuring, mainly driven by administrative and state enterprise reforms that started in 1998 and focus on separating administrative functions (government functions) from actual operations (business management functions). In the past, government organizations were deeply involved in such aspects as personnel matters, investment decisions and asset disposal of the state-owned enterprises. Due to this influence, especially small-scale investments were sometimes made irrespective of economic efficiency or market demand. In some projects, vast funds were borrowed and promotions were not always based on merit, subsequently leading to inefficient management. This is changing for the better. In China about 54,000 chemical sites are said to be in place. These companies are either collectively owned (provincial or municipal ownership) or, increasingly, privately owned. Some of the collectively owned enterprises are spin-offs from Sinopec and PetroChina following their restructuring in 1998.

Management is an obvious weakness of Chinese enterprises. Reform efforts with the introduction of corporate governance have just begun. It will take time to catch up with international standards in this regard. Redundancy remains serious despite massive lay-offs. The number of employees is still at least five times larger than the world average.

Estimates show that compared with the average prices of ethylene products from neighboring countries like Japan and ROK, domestic prices are 20 to 30 percent higher. This difference includes 8 to 12 points in manufacturing costs, 3 to 5 points in financial costs, 1 to 2 points in management costs and 3 to 8 points in tax costs. China can no longer afford small ethylene units.

The overall conclusion is that China will need to invest in substantial refining and petrochemical plant capacity in the coming decade. It will be critical for China's long term competitiveness in the global marketplace to ensure that these new plants are world-scale and world-class, and capable of competing with the best facilities anywhere. China has decided to:

- close small and inefficient plants;
- cut operating costs;
- reduce redundancy;
- restructure institutions;
- upgrade technologies and equipment;
- integrate refineries and petrochemical plants with sales and distribution companies;
- expand the sales network;
- strengthen international cooperation.

The bottom line is that the continuing strong growth of the Chinese economy will spur imports of crude oil, petroleum products and petrochemicals, despite massive and ambitious investment plans within China. Imported crude oil will become increasingly heavy and sour, requiring new refining configurations and modification of existing plants. The product shortfall will grow over time and will be concentrated in the south and southeast. As for chemicals, there will be a substantial continuing dependence on imports, despite an aggressive debottlenecking and grassroots construction program.

The investment implications of this outlook are quite challenging, with the petroleum and petrochemical industry in China facing a huge task. The question is not whether these investments are needed, but rather whether the technical, managerial and financial capabilities can be marshaled to safely and efficiently develop competitive world-class facilities.

2.8 Summary and Outlook

The development of China's petroleum and petrochemical industry has gone through massive changes, as it was struggling to stay in sync with the fast growing Chinese economy.

This capital and know how intensive sector has developed from a small sized regionally diverse industry into an increasingly globally competitive industry, partly also due to the heavy engagement of multinational companies. China's raw material needs especially for oil make china internationally more interdependent, and the secure supply of oil determines the industries future. Refineries are debottlenecked, new capacities come on the market. The fast growing economy with China being the future production center of the world requires a competitive and reliable chemical industry, and China focuses on new crackers and the expansion of existing facilities, to serve as an industry of the industry. China has the fastest growing chemical market in the world, and the focus is on synthetic resins, synthetic rubber and synthetic fibers. China's entry into WTO has altered the economic and legal landscape, opened up new possibilities for investors as well as importers. The competitive pressure in this market is very high, and only the respective market leaders with world scale facilities will prevail.

Given the fact that China is already the second largest economy in purchasing power parity terms (in 2004 7th largest economy in real GDP terms) and about to grow its economy by 6-7% over the next decade, this determines the growth pattern of China's petroleum and petrochemical industry. China will have the world's largest automobile industry in 2030; half of its population by then will be a middle class with a consumption pattern that reflects that of Southern Europe, and its energy needs and chemical demand will reshape the global markets. Not only will China consume more and more oil, gas, steel, iron ore and copper, but will have to develop a sustainable economy to avoid the environment and the international commodity markets to collapse. Modern technologies in the chemical sector will help China to grow without outgrowing the system. The 21st century looks set to become the Asian Pacific century, and the European companies are already in China to become part of this unprecedented growth story.

3 Activities of European Chemical Companies in China

Heinz Mueller: DZ Bank AG, Frankfurt, Germany

3.1 Executive Summary

The Chinese economy has been growing at an average annual rate of 9 percent for the last ten years. There is no apparent end to this expansion. The chemical industry's ambition is to participate in this economic growth, so we can see two trends in the chemicals industry. Firstly, as an upstream supplier industry, the chemicals players are obliged to go along with their customers and also invest locally. If they don't, they can expect to lose their customers to local competitors. Secondly, their lower return requirements are enabling Chinese competitors to undercut European manufacturers on the prices of their products. In the case of a handful of chemical products already, the export of made-in-China goods to Europe is now putting the European chemicals sector under pressure at home.

We have investigated how the European chemical companies we analyze in our equity universe are responding to these challenges. We have identified two factors that are likely to be relevant to the future success of our chemical companies: the "first mover effect" and a "focus on subsegments". The "first mover effect" reveals which chemical companies have moved quickly enough to establish their positioning in China in time. We identify this by investigating their China exposure and investment plans. The "focus on subsegments" factor shows which companies are active in the fastest growing areas of the market. We accordingly analyzed the positioning of our population of chemical companies in China and their growth prospects through to 2008. We used a score card analysis to weight each company's results on these success factors relative to their peers and generate indications of their chances of success in China.

The chemicals market in China is expected to grow at a CAGR (2002 to 2015) of 5.4 percent – much faster than the world market (2.7 percent). Many of the chemicals sector's most important customer industries have spent billions building enormous production capacity in China. Hence, the chemical companies have no choice but also to produce locally in China in order to retain their customers' loyalty through the timely and low-cost provision of chemical products.

The demand for basic chemicals is likely to remain strong. Chemical companies that are focused on this lower end of the value chain are likely to benefit most from China's upswing. Companies that manufacture products at the top end of the value chain (specialty chemicals) are more likely to be later-stage beneficiaries of the China boom, when there will be more demand for higher-value products.

3.2 Activities in China

There are significant differences between our selected companies with respect to their product portfolios, strategies and positioning in China. To reflect these we designed a score card that helps analyze the chemical companies by reference to the two key success factors we have identified: the "first mover effect" and "focus on subsegments". The "first mover effect" factor identifies the chemical companies that have got the timing of their China positioning right. We judge this by two criteria: firstly their China exposure, in other words the companies' China turnover compared with their total group sales; and secondly the companies' planned investment in China. The "focus on subsegments" factor identifies which companies are exposed to the fastest-growing markets in China. We also measured this by analyzing two criteria: firstly the chemical companies' positioning in China and secondly their growth prospects in China through to 2008. We entered the four criteria derived from our two critical success factors as categories in a score card table and then compared the chemical companies. The chemical companies naturally differ in their performance on the four criteria. We awarded points (between zero and four) depending on the degree to which they satisfied our defined criteria. The more fully each criterion was met, the more points a company could gain (up to a maximum four). We then entered the average of each company's score in the four categories in the right-hand column of the table. This represents our qualitative assessment of the companies under investigation. The higher the points total on the right of the table, the better the chances in our view that the relevant China strategy will pay off by 2008.

In the "China 2003 share of sales" category we compare the China sales of each chemical company in fiscal 2003 with its global sales figure. The bigger the quotient, the more important its China operations are for the company concerned. Of the chemical companies analyzed, Fuchs Petrolub came out at the top of this category. The group generated 2003 sales of EUR 57.2 million in China, equivalent to 5.5 percent of group turnover. At 4.8 percent BASF has the second biggest China exposure, but of course it is ahead in absolute terms (EUR 1.6 billion). DSM (4.5 percent) is only a little behind these two leaders. We accordingly awarded all three companies the full score of four points. Bayer and Akzo Nobel come next with China exposures of 3.9 and 3.7 percent, and we awarded these companies three points in this category. At 2.5 percent, Degussa lags behind the other chemical companies somewhat and only got two points from us. The agrochemicals companies have the lowest China exposure, so we gave K+S (2.0 percent) and Syngenta (1.5 percent) one point each and KWS (0.5 percent) no points.

In the "investment plans in China" category we investigated the extent to which the chemical companies have made or are planning to make investments in China (*Table 3.1*). The bigger the scale of the investment and the earlier it was initiated, the better the companies meet this criterion. BASF came out on top in the "investment plans in China" category. BASF has laid the foundation for a massive expansion of its China activities through its 50-percent stake in the US\$ 3 billion investment in a vertically and horizontally integrated plant in Nanjing that got under way ten years ago, as well as its major participation in the US\$ 1.3 billion isocyanates production joint venture in Caojing. It follows that BASF earns the maximum points (four) in this category. Bayer has also budgeted for big investment in China (US\$ 3.1 billion in Caojing). However, Bayer has had to put back the completion date for its planned MDI/TDI plant from 2006 to 2008 or 2009. This shows that Bayer was relatively late with the timing of its investment (November 2001). This is a black mark, which is why we have only given Bayer three points in this category. K+S and KWS have no real ambitions to make significant investments in China. They accordingly get no points. The remaining companies intend to make investments in the double-digit millions (Syngenta) or low hundreds of millions of euros range (Degussa) in China and therefore get either one or two points.

Table 3.1. Qualitative evaluation of chemical companies in China

Company	"First mover effect"		"Focus on subsegments"		Chance of strategy succeeding by 2008e
	Share of 2003 sales	Investment plans	Position-ing	Growth out-look to 2008e	
Akzo Nobel	•••	•	••	••	••
BASF	••••	••••	••••	••••	••••
Bayer	•••	•••	•••	•••	•••
Degussa	••	••	••	••	••
DSM	••••	••	••	•••	•••
Fuchs Petrolub	••••	••	••••	••••	••••
K+S	•		•	•	•
Syngenta	•	••	••	•	••

The "positioning" category looks into the chemicals subsectors in which our companies have chosen to become active in China. Since an emerging country such as China will naturally have especially strong demand for products from lower down the value chain (e.g. basic chemicals) at the present juncture, basic chemicals suppliers have a particular advantage. BASF and Fuchs Petrolub stand out in this category with scores of four points each. Following the completion of its cracker plant in Nanjing at the end of 2004, basic chemicals will probably account for around 20 percent of BASF's China-generated portfolio. The demand for these commodity chemicals is expanding at double-digit rates and it will probably be many years before it can be met from local production. BASF is the biggest beneficiary from this situation. Like BASF, Fuchs Petrolub is active at the lower end of the chemicals production chain. However, Fuchs lubricants are specialty products. They are in hot demand in China from the booming steel (especially metal refining and metalworking) and automotive industries. While car sales in China increased by around 50 percent in 2003 to 1.8 million vehicles, J.D. Power predicts annual market growth of 20 percent in the next few years. This means that Fuchs Petrolub is also well positioned to participate in China's chemicals boom. Bayer, Degussa, DSM and also Akzo Nobel intend to increasingly prioritize specialty chemicals in the next few years. A survey of the chemical companies we conducted showed that specialty chemicals should enjoy above-trend growth in China in the coming years. However, we only see this subsector truly blossoming after 2005. The companies that are focusing on this sector in China therefore score fewer points. While Degussa says the fine chemicals for pharmaceutical products sector in China is predicted to grow by between 8 and 13 percent annually from US$ 10 billion at present to US$ 15 billion in 2010, the global market is likely to grow at a slower, single-digit rate. The some 5,000 Chinese fine chemicals companies are already taking market share in Europe from the domestic companies through aggressive price competition. We therefore see these companies' fine chemicals exposure as a negative factor. Within our coverage, DSM and Degussa in particular face this threat. Both companies accordingly only get two points in this category, despite their specialty chemicals exposure. The agrochemicals companies get either one (KWS, K+S) or two points (Syngenta), as they are more likely to benefit in the long term from China's increasing prosperity and the resulting changes in people's dietary habits.

Under the "growth prospects through to 2008" category we estimated the China sales growth rates of our population of chemical companies from 2003 through to 2008. The higher the growth rate, the better they met this criterion. In this category we see BASF (estimated CAGR 2003 to 2008: 17.4 percent) and Fuchs Petrolub (estimated CAGR 2003 to 2008: 16.0 percent) at the top of the ranking, so both score maximum points again. We see the agrochemicals companies enjoying the lowest average annual growth through to 2008 (Syngenta: 7.2 percent; KWS Saat: 6.8 percent; K+S: 6.5 percent), and award them just one point each. The other players get three points (Bayer: 14.2 percent; DSM: 13.5 percent) or two points (Akzo Nobel: 10.0 percent; Degussa: 9.5 percent) depending on their growth rates.

The average points for each company across all four categories can be read off from the right-hand column of our score card. This shows that BASF and Fuchs Petrolub merit the most points, four each. It is above all their strong positioning in China, their exceptional exposure and their high future growth rates that lead us to conclude that these two companies have the strongest chances of succeeding in China. The agrochemicals companies Syngenta and KWS especially, but also to a lesser extent K+S, have come off worse in this comparison than the other chemical companies. In the long term, which means beyond our detailed planning horizon of 2008, the agricultural sector in China could become more important for the chemicals industry. This is because increasing prosperity is likely to change people's eating habits (e.g. more meat) in China, and this is likely to require more capital-intensive farming and thereby benefit the agrochemicals companies.

3.3 Investment in China

Trend: The Chemical Industry's Customers Are Themselves Increasingly Producing in China

While it is true that European chemical companies can always profit from the China boom by exporting, a trend is currently emerging that dramatically changes the situation for the developed countries' chemical companies. Many of these companies' customers have already opened production facilities in China and they are actively planning to add further local production plants. They have invested several billion euros to date and are committed to investing further billions in China. Companies in the automotive sector are some of the chemical industry's most important customers. This means that the chemical companies wanting to supply these automotive manufacturers now need to establish an even stronger presence in the Chinese market if they are to continue to market their products there and expand their operations. At present, German chemical companies in the Asia Pacific region generate no more than 58 percent of their revenues on average from local production. The ratio for China is likely to be even less than 50 percent. The implications are firstly that China will continue to import chemicals for years to come, and secondly that the chemical companies can safely be expected to continue to invest heavily in China in the future.

However, getting into the Chinese market is not a simple proposition. This is why it makes sense for foreign companies to work with a domestic partner that has the necessary contacts. Where foreign companies wish to produce in China in strategically important sectors of the economy (such as chemicals production), it is even a statutory requirement for them to enter into a joint venture with a Chinese partner. They are however allowed to handle the subsequent distribution through majority-owned subsidiaries.

Producing chemicals in China has the advantage for foreign chemical companies of allowing them to supply their demanding customers, such as the automotive industry, on a fast-response basis – an absolute must for any supplier in these days of just-in-time production. The rapid growth of China's market is also driving up demand from consumer industries substantially (automotive, building, electronics etc.). Their needs for chemicals can barely be satisfied locally, so their production facilities are always likely to be working at full capacity, which helps reduce their fixed unit costs.

On top of all this, wage costs are much cheaper in China than in Western Europe. Wages in China are one twentieth of those in the United States or Germany. While an hour's labor in the USA costs an average of EUR 16.14, in China it only costs the equivalent of around 60 US cents. Admittedly, labor efficiency in China has not reached western standards yet. Chemical companies reckon that to produce in China they need three times the labor compared with Germany. A Degussa survey showed that personnel costs in the chemicals industry in China are only one third of what companies have to pay for labor in Western Europe. Despite the fact that the chemicals industry is a capital-intensive sector of the economy, its wage bill is still substantial. Switching from exporting to local production could be a way to cut manufacturing costs and boost profitability. Local production also eliminates the costs of transporting exports as well as steep import duties. Local production is therefore a potential source of competitive advantage.

Trend: Chinese Fine Chemicals Producers Are Presenting Europe with Stiff Competition

Chinese chemical companies are certainly making the most of these competitive advantages, especially in the fine chemicals subsegment that manufactures intermediate products for the pharmaceuticals industry. Producing fine chemicals in China is significantly cheaper than in the developed regions. The costs of producing fine chemicals in China are only 40 percent of European or U.S. costs. This cost advantage is only slightly offset by the cost of transporting the goods to Europe. This is why large numbers of Chinese fine chemicals companies are able to export their products to Europe at a very favorable price. This is inflaming an already aggressive price war in Europe. The result however is that European fine chemicals suppliers also have to make substantial price concessions to stay in business in Europe. European fine chemicals producers are therefore under severe pressure. This situation is especially threatening since most European chemical companies do most of their business in Europe.

Investment in China Is a Must

Chemical companies will therefore increasingly have no choice but to open more of their own production facilities in China in order to meet their customers' requirements for rapid and low-cost supplies. Chemical companies that are already producing locally are therefore likely to benefit the most from the strong growth of the chemicals market in China (*Table 3.2*).

Table 3.2. Selected chemical companies' activities in Greater China (Source: Companies, DZ Bank estimates)

Company	Sales in China (2003)	China sales as % of total sales	Total invested in China	Investment segments	Anticipated production quantity p.a.	Anticipated production start
Akzo Nobel	EUR 510 m	3.9%	Double-digit millions of euros	Coating powder production Coatings, chemicals, pharma	n.a.	2004 2004
BASF	EUR 1.6 bn	4.8%	US$ 1,000 m US$ 300 m US$ 2,900 m	Production JV for MDI and TDI isocyanates THF/PolyTHF JV for ethylene and integrated chemical site total	MDI: 240 Tt TDI: 160 Tt 80/60 Tt 600 Tt 1,700 Tt	2006 2005 2005 2005
Bayer	EUR 1.1 bn	3.9%	US$ 3,100 m in total US$ 1,100 m US$ 450 m	Coatings production MDI TDI Polycarbonate production (Makrolon)	50-70 Tt 230 Tt 160 Tt 100-200 Tt	2004 2008 2009 2006
De-gussa	EUR 280 m	2.5%	> EUR 100 m	Carbon black, Pharma amino acids Polyurethane foam production	70 Tt	End 2004
DSM	EUR 273 m	4.5%	n.a.	Semi-synthetic cephalosporin food products (nutritionals) Food premixes Engineering plastics Unsaturated polyester resins Synthetic fibers and plastics	140 Tt	2005
Fuchs Petro-lub	EUR 57 m	5.5%	Low millions of euros	Metal surface lubricants		
K+S	EUR 36 m	1.6%	0	Nitrogen fertilizers	0	-
KWS	EUR 2 m (e)	0.5%e	0	Sugar beet seeds Sunflower seeds Oil crop seeds Corn seeds	0	
Syn-genta	US$ 75m (e)	1.1%e	US$ 85 m	Non-selective herbicide (Gramoxone)	6 Tt	2001

BASF was one of the first chemical sector companies to begin investing billions in China. It was as far back as 1994 that BASF began the planning of its fully integrated plant in Nanjing, which it will operate as a joint venture with its Chinese partner Sinopec. This early planning is now paying off and leaves BASF in the fortunate situation of being at the moment the only local producer of the much sought-after ethylene. We think that BASF should benefit from this situation. The group's China sales already amount to EUR 1.6 billion.

Bayer was the first of all the companies in our coverage to be involved in China. It made its first sales in the country back in 1882. Bayer currently has significant sales in Greater China of EUR 1.1 billion. This works out at 3.9 percent of total group sales. To gain access to the Chinese market without creating excess capacity, Bayer has negotiated a contract for BASF to supply it with raw MDI in China. Bayer markets MDI separately under its own brand and this is helping to create a competitive market.

The other chemical companies in our coverage are planning to invest more moderate or even small sums in China. Degussa stands out a little from the crowd in that it is planning investments totaling more than EUR 100 million in China. Although the company has been active in China since 1933 and has been producing in China since 1988, Degussa's local sales of EUR 280 million only give it a China exposure of 2.5 percent. Akzo Nobel has a moderate footprint in China. Its China sales are around EUR 510 million, which as a proportion of total group sales is 3.9 percent. It has only set aside a double-digit million euro amount for investment. Syngenta, which turns over an estimated US$ 75 million in China or around 1.1 percent of group sales, intends to invest US$ 85 million in China. Fuchs Petrolub has the biggest China exposure at 5.5 percent of total sales (China sales: EUR 57.2 million) and is planning to invest a "low million euro" sum in the country. Some companies such as DSM, whose China sales (EUR 273 million) account for 4.5 percent of group turnover, will not reveal the size of their investment plans. Others such as K+S (China exposure: 1.6 percent) or KWS (China exposure: 0.5 percent) are not planning any investment.

3.4 Focus on Subsegments

Growth Varies Between Chemical Sector Subsegments

Our analysis has differentiated between the subsectors of the chemicals industry because we expect their growth patterns to vary significantly. China will remain a future net importer of raw materials especially (e.g. crude oil) and most importantly of petrochemicals (ethylene, propylene). The known investment projects in the pipeline indicate that the currently planned capacity expansion is unlikely to be able to cover the predicted demand. Petrochemicals growth is therefore likely to remain strong. This means that chemical companies that are focused on this lower-end range of the chemical production chain are likely to profit from this strong growth. China's low production costs represent an enormous cost advantage over chemical facilities in the established economic regions. This means that future Chinese exports represent a serious potential threat to chemical companies that produce in the industrialized countries. The market for vitamins is now serviced almost exclusively from China. Supplies of fine chemical intermediates for the pharmaceuticals industry are increasingly also sourced from China. The country is now a feared competitor as a manufacturer of fine chemicals for pharmaceutical industry applications – not least because of the price squeeze it has initiated.

Basic Chemicals Currently Comprise the Most Important Subsegment of China's Chemicals Market

In a study of the chemicals market in China in 2002, the management consultancy A.T. Kearney said the market could be divided into several subsegments: basic chemicals (58 percent), fine chemicals for pharmaceuticals (15 percent) and specialty chemicals (11 percent). We have further divided the fourth A.T. Kearney category - "consumer chemicals" (16 percent) - into two subsegments: consumer chemicals (10 percent) and agrochemicals (6 percent) to reflect the subsequent changes in market relativities (*Fig. 3.1*).

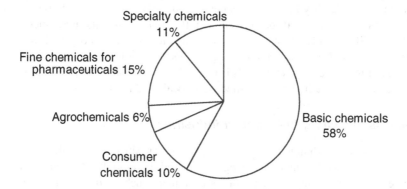

Fig. 3.1. Segmentation of the chemicals market in China in 2003 (Sources: A.T. Kearney, Frost & Sullivan, Henkel, DZ Bank Research)

By consumer chemicals we mean detergents and cleaning agents plus cosmetics and personal care products. Agrochemicals covers crop protection products (herbicides, insecticides and fungicides), fertilizers (nitrogen, phosphates and potash) and seeds. Specialty chemicals primarily cover (high-grade) plastics. Fine chemicals for pharmaceuticals comprise the manufacture of primary and intermediate products for the pharmaceuticals industry. Basic chemicals comprises substances such as ethylene and propylene. The recent years of market growth in China have been dominated by meeting the demand for these basic chemicals, which is why this segment accounts for by far the biggest relative share of the Chinese chemicals market as a whole.

China is currently changing from being a net exporter of primary materials into a net importer, e.g. crude oil. The same thing can be seen happening in basic chemicals, the raw materials for chemical products. China's enormous economic growth is generating demand for massive quantities of petrochemical products and starter chemicals. This will be the biggest growth area in the next three or four years and will outpace the chemicals market as a whole. Basic chemicals are produced in so-called cracker plants that require an investment of billions of euros. Since China is determined to establish a presence in the basic chemicals segment, it is assured of very high priority status. Projects of this kind are mostly built and subsequently operated on a joint venture basis involving a Western and a Chinese company. A production facility can take years to build, especially if one counts the planning phase and the time it takes to get the necessary licenses. The BASF, Shell and BP crackers that are currently under construction probably already have a ten-year planning phase behind them. Since we know of no other similar production plants in the pipeline in China apart from these three crackers, they are likely to be a very important source for supplying China's chemical industry with starter chemicals over the next ten years.

The construction of the cracker plant in Nanjing will allow BASF to expand its product portfolio in China and increase the weighting of basic chemicals. Once the integrated facility in Nanjing is completed, we estimate around 20 percent of BASF's China sales will be derived from the basic chemicals subsegment. This makes BASF the only company in our coverage that is focused on this lower end of the value chain and will therefore get a boost from the basic chemicals boom. Demand for these basic chemicals is expanding at double-digit rates and it will be many years before it can be satisfied from local production.

The Focus Is Switching to Customer Industries

Structures are bound to change in the long term however, since the chemicals industry is a supplier to other manufacturing sectors and therefore has to change as its customer structure changes. Many of its customers such as the automotive, building and electrical/electronics industries are investing billions to build their own production plants in China. While the main focus in China so far has been on supplying primary chemical products to other chemical companies, products for other industries are set to become more and more significant in future. This means chemical companies will increasingly be offering higher-value products from further up the value chain. The so-called specialty chemical products are occupying an increasingly important place in the chemical companies' product ranges. This is likely to have a major impact on the structure of the chemical sector in China.

We expect the structure of the chemical sector in China to change substantially over the course of the next 11 years. The management consultancy A.T. Kearney has predicted that the relative share of basic chemicals could decline from 58 to 40 percent in China over this period. It follows that the other segments will gain relatively in weight. A.T. Kearney believes specialty chemicals could experience the biggest expansion of any segment. Specialty chemicals' share of the chemicals market could accordingly rise from 11 to 20 percent. Fine chemicals could expand from 15 to 18 percent. Consumer chemicals could widen its share from 10 to 15 percent, and agrochemicals should increase from 6 to 7 percent (*Fig. 3.2*).

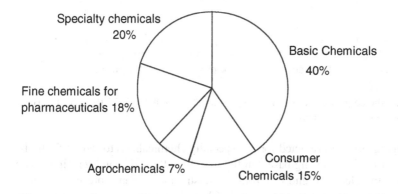

Fig. 3.2. Expected segmentation of the chemicals market in China in 2015 (Source: A.T. Kearney, DZ Bank)

Specialty Chemicals Will Enjoy Above-Trend Growth

To establish the long-term growth prospects for the various chemical industry subsegments in China, we conducted a survey of several chemical companies and asked how they see the subsegments evolving over the period from 2002 through 2015. Before we discuss the results of the survey, we have to point out two shortcomings. Firstly our respondents were not prepared to comment on the consumer chemicals sector, and we therefore have no estimates for this sector. Secondly, since some subsegments will grow faster than the market, there should by definition also be segments that grow proportionately slower. The survey has not revealed which they are, however. The reported growth rates for the individual chemical sectors therefore need to be taken with a pinch of salt. The differentiation of the growth rates for the individual chemical market subsegments shown in the chart below should however serve to illustrate the general direction of how the subsegments will probably develop relative to each other. This shows that specialty chemicals is expected to grow the fastest. Its growth rate should be much higher than the average growth of the chemical market in China. Basic chemicals should also grow at above-average rates, as should fine chemicals, though the latter will lag behind specialty and basic chemicals quite clearly. The companies surveyed predicted average growth for the agrochemicals sector in China (*Fig. 3.3*).

Growth rates

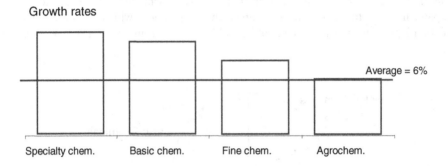

Specialty chem.   Basic chem.   Fine chem.   Agrochem.

Fig. 3.3. Predicted growth of chemical sectors in China, 2002 – 2015 (Source: DZ Bank survey of chemical companies)

The strong growth predicted for the specialty chemicals sector could firstly be a product of the low starting base. On the other hand, the demand for high-grade plastics for the electrical and electronics industries and even more for the auto-motive sector is very strong. Auto sales in China for instance increased by around 50 percent in 2003 to 1.8 million vehicles. J.D. Power predicts sales will grow annually by 20 percent in the next few years. This means that chemical compa-nies that manufacture primary products for these industries stand to profit dis-proportionately in the long term from the growth of China's chemicals market. The biggest suppliers to the automotive and electronics industries in our coverage are BASF, Bayer and Degussa, plus Akzo Nobel and DSM. It follows that these groups should be beneficiaries of the strong long-term growth in the specialty chemicals subsegment. We see Fuchs Petrolub as especially interestingly posi-tioned. As a lubricants producer, Fuchs Petrolub stands to gain from China's booming demand for automobiles.

Most Fine Chemicals Growth Will Be in China

On the other hand, fine chemicals, most importantly for pharmaceutical applica-tions – are currently suffering from excess global capacities. The players' earlier hopes that the pharmaceuticals industry would outsource its production to them have failed to materialize. Instead, the wave of mergers in the pharmaceuticals industry is causing them problems as production is concentrated to release syner-gies. The resulting (excess) capacity is just being released. What we are seeing instead is a growing insourcing trend that is hitting the fine chemicals market. While the global fine chemicals market for pharmaceuticals is set for at best low single-figure growth, Degussa predicts that the pharmaceuticals fine chemicals sector in China will expand by between 8 and 13 percent each year from US$ 10 billion at present to US$ 15 billion in 2010. The reason for this strong fine chemicals growth in China is that the cost of producing fine chemicals in China is only 40 percent of the equivalent U.S. production cost. It is no surprise therefore that the profitability of Chinese fine chemicals companies is soaring and has already caught up with the European and American level. Their low costs are allowing the Chinese fine chemical companies to increasingly win orders from big multinational pharmaceutical players. They not only supply their home market but are also breaking into Europe with aggressively priced exports. Competition is increasingly being fought out through prices. European fine chemicals companies have to fear for their market shares in their home region as their customers substitute cheap Chinese products for expensive European ones. We see the European companies that are heavily exposed to fine chemicals coming under increasing pressure. Within our coverage it is first and foremost DSM and Degussa, but also Akzo Nobel, which are facing this threat from the Far East. The European fine chemicals companies will therefore also be obliged to produce cheaply in China if they want to survive the competition.

Good Long-Term Prospects for Cosmetics and Detergents

The consumer chemicals sector (manufacturing of detergents and cleaning materials plus personal care and cosmetic products) is also likely to grow disproportionately fast in China. The manufacturers offering such products will profit from China's annual population growth of almost 0.7 percent. In the long term, increasing affluence will also boost the demand for consumption-related goods. This in turn will boost the demand for cosmetics and personal care products, as well as for washing powders and cleaners. While personal care and cosmetic products will be driven by the desire to look and feel better, it is the convenience character (comfort factor) of detergents and cleaning agents that is the decisive driving force.

Growth in Agrochemicals Will Be Mainly Long-Term

The agrochemicals sector (crop protection products, seeds, fertilizers) probably offers mainly long-term potential. Rising prosperity also tends to increase the consumption of meat (eating habits change). This requires more intensive animal production and therefore higher crop yields that can be achieved by the use of crop protection products and fertilizers. Increased quantities of animal feedstuffs are also required. It follows that chemical companies that supply the food and feedstock industries will also have a chance to profit from population growth and rising affluence. Crop protection manufacturers will also be able profit in the long term from China's expanding market. Rising wages are also likely to make the use of crop protection products increasingly worthwhile in China, especially for the two most important agricultural products: cotton and rice. This could boost companies such as Syngenta and Bayer. To a slightly lesser extent, we also see BASF, KWS and K+S as beneficiaries of this trend. Since change takes a very long time in the crop protection, fertilizer and seed markets, this trend is likely to have less relevance to these chemical companies in a short-term view.

Investors Attach Greater Weight to Short-Term Prospects

The structural changes we predict for the chemicals industry in China have very different implications for the individual chemical sector subsegments (*Table 3.3*). However, stock markets attach more weight to short-term upside prospects than to long-term promise. This is why we regard the short-term structural shift in favor of basic chemicals as probably more relevant to chemical companies than the long-term shift in favor of specialty chemicals. We therefore attach more weight to the basic chemicals trend than to the specialty chemicals trend. Chemical companies that are active in the basic chemicals subsegment and have already built a good foundation in China should accordingly do better from the China boom than chemical companies belonging to the other segments.

Table 3.3. Portfolio structure in China, 2003 (Source: DZ Bank estimates)

Company	Basic chemicals	Specialty chemicals	Fine chemicals	Agro- chemicals	Consumer chemicals	Pharma
Akzo Nobel	0%	70%	0%	0%	0%	30%
BASF	0%	90%	5%	5%	0%	0%
Bayer	0%	70%	5%	5%	0%	20%
Degussa	0%	90%	10%	0%	0%	0%
DSM	0%	80%	20%	0%	0%	0%
Fuchs Petrolub	0%	100%	0%	0%	0%	0%
K+S	0%	0%	0%	100%	0%	0%
KWS	0%	0%	0%	100%	0%	0%
Syngenta	0%	0%	0%	100%	0%	0%

- Basic chemicals: ethylene, propylene etc.
- Specialty chemicals: incl. plastics, nutritionals
- Fine chemicals : primary and intermediate products for pharma
- Agrochemicals: fertilizers, pesticides, seeds
- Consumer chemicals: cosmetics & personal care; detergents & cleaning agents
- Pharma: incl. OTC

None of the companies is actually doing basic chemicals business in China as yet. Even BASF, the only company in our coverage that is a force in this subsegment, has not started operations in China in this field so far. But once the construction of its cracker in Nanjing is completed in the fourth quarter of 2004 and the entire plant comes on stream from early 2005, basic chemicals are likely to generate as much as 20 percent of BASF's China sales. It follows that BASF is the only company with an opportunity to start to grab some of the action in the fast-growing base chemicals market even in the near future. In the long term, the specialty chemicals suppliers will also profit from the China boom. Since Fuchs Petrolub is 100-percent exposed to specialty chemicals in China, we see this company as the best positioned of the specialty chemicals companies.

3.5 Effects of Investment and Positioning on Sales

Based on the companies' current sales in China, their degree of establishment, their published investment plans and the growth prospects for the various subsegments, we have estimated the effects of their China activities on each of our population of chemical companies. We have then used these figures to identify which of the chemical companies in our coverage stands to profit the most in future from the China boom (*Table 3.4, Fig. 3.4*).

Table 3.4. Anticipated business growth of selected chemical companies in China, 2003 to 2008 (Source: Company data, DZ BANK estimates)

Company	China sales 2003 (m)	Currency	China sales 2008e (m)	CAGR (03-08e)	China as share of total sales 2003	China as share of total sales 2008e
Fuchs Petrolub Vz.	57.2	EUR	120.1	16.0%	5.5%	10.0%
BASF	1,600.0	EUR	3,564.2	17.4%	4.8%	9.3%
Bayer	1,100.0	EUR	2,135.2	14.2%	3.9%	6.3%
DSM*	273.0	EUR	514.3	13.5%	4.5%	6.2%
Akzo Nobel	510.0	EUR	821.4	10.0%	3.9%	6.1%
Degussa	280.0	EUR	440.8	9.5%	2.5%	3.5%
K+S	36.0	EUR	49.3	6.5%	1.6%	1.8%
Syngenta	75.0	US$	106.2	7.2%	1.1%	1.4%
KWS SAAT AG	2.0	EUR	2.8	6.8%	0.5%	0.5%

*Acquisition of Roche's vitamins business on 1.10.2003: Annual China sales EUR 100 million

As of group
sales 2008e

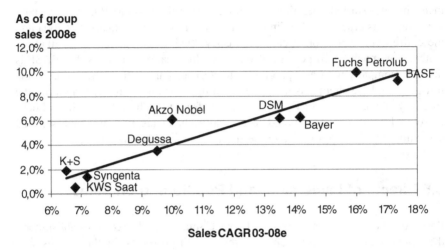

Fig. 3.4. Development of China activities (Source: DZ Bank)

Within our coverage Fuchs Petrolub is the chemical company that generated the highest proportion of its 2003 turnover in China. Its China sales accounted for 5.5 percent of group turnover. Its focus on specialty products is paying off. The company is profiting from China's fast-growing auto market and booming steel industry. The growth of the lubricants market in China suggests Fuchs will double its sales within five years. China is therefore likely to account for an estimated 10.0 percent of group turnover in 2008, the highest proportional China component in our coverage.

BASF's China sales of EUR 1.6 billion are currently (2003) the highest in absolute terms of all the chemical companies in our coverage. This is equivalent to 4.8 percent of group sales worldwide. Because of the massive investment in BASF's integrated site ("Verbund") in Nanjing and the Caojing production facility, and assuming that local production operates at full capacity and chemicals also continue to be exported, we expect BASF to enjoy the biggest turnover growth. We predict a CAGR from 2003 to 2008 of 17.4 percent, which is the highest growth rate within our coverage. Both Fuchs Petrolub and BASF are likely to be achieving by far the highest percentages of their total turnover in China in 2008 compared with the rest of our population.

Bayer will also benefit from China's strong chemicals growth. However, since the company only laid the foundation stone for its billion euro China investment in November 2001, we do not expect the sales of its polycarbonate and polyurethane products to produce major growth before 2006 onwards. We estimate Bayer's average growth rate for the next five years at 14.2 percent. DSM also appears to benefit on the strength of its high growth rate of 13.5 percent and anticipated China sales share of 6.2 percent in 2008. The company aims to double its China sales to EUR 600 million by 2008.

Degussa is likely to probably grow in line with the market. We see a 9.5 percent average rate of increase through 2008. Its planned total investment of more than EUR 100 million looks minor alongside BASF's and Bayer's activities. Akzo Nobel is planning to invest even less. Thanks to its significant China exposure and good positioning in coatings, we predict the CAGR between 2003 and 2008 of Akzo Nobel's China business at 10.0 percent. Chemical companies like Syngenta, KWS and K+S will probably only be able to marginally raise their China sales' relative share of total sales. The main reason will be that these companies are either not investing in China at all, or only on a very small scale. China is not a priority for them.

3.6 Selected Chemical Companies in China

Many chemical companies have recently been stressing the opportunities that China offers the chemicals industry. While Akzo Nobel, BASF, Bayer, Degussa, DSM and Fuchs Petrolub are consciously focusing on China, China is not a priority for K+S, KWS and Syngenta. However, all the companies' business activities have different features and emphases (*Tables 3.5 and 3.6*).

Akzo Nobel

Akzo Nobel has been active in China since the early 1980s. The company's 2,500 strong workforce generated 2003 sales in China of EUR 510 million (EUR 435 million in the People's Republic of China; EUR 75 million in Taiwan). This was probably broadly the same as the preceding year. These sales cover all three of the group's business segments (pharmaceuticals, coatings, chemicals). The coatings segment accounted for almost two thirds of the total. The group's China exposure is 3.9 percent. The group is planning investments in the double-digit millions of euros, which is a fairly moderate commitment compared with the other chemical companies in our coverage. The investment focus is on the coatings segment (paints and lacquers), which we count as part of the specialty chemicals subsegment. Akzo Nobel is determined to expand its leadership in powder coatings. Its decorating paints and auto lacquers are allowing the company to cash in on the strong demand from the building and automotive industries. Its 21 plants in China mean that 70 percent of its sales are locally produced. This is allowing the company to build up strong customer loyalty, another positive trend. We estimate a sales CAGR between 2003 and 2008 of 10.0 percent for Akzo Nobel in China.

Table 3.5. The chemical companies' activities in China (Source: Company disclosures)

Company	Nature of activity and business model	Current products in China	Strategy	Positioning
Akzo Nobel	Trading in all three segments; more than 20 production plants	Coating powders, other coatings, dyes	To expand leading role in coating powders through local production; investment in double-digit millions planned	Leader in powder coatings
BASF	Trading in four out of five segments; oil and gas the exception; both production sites (12) and sales offices (18); BASF has ten wholly owned subsidiaries and 9 JVs	Chemicals, performance products, plastics, agricultural products and nutrition	Focus on ethylene and plastics production; China destination for two thirds of Asian investment in the years 2000 to 2005. Invests with partners US$ 2.9 billion in Nanjing until 2005 and EUR US$ 1.3 billion in Caojing until 2006	One of biggest international players in China, but still on small side compared with local players
Bayer	Trading in all three segments; both production and sales subsidiaries	Rubber chemicals, leather chemicals, pharmaceuticals (Canesten, Talcid, Glucobay and Nimotop), crop protection products, pigments, polycarbonate intermediates, veterinary medicines, coating raw materials, polyurethane production systems	Focus on polymers production (polycarbonate, coating raw materials incl. starter materials, raw materials for polyurethane production); start made in Nov. 2001 on biggest investment project in Caojing (total US$ 3.1 billion)	Strong position in polymers; therefore investment focused mainly on this segment where local producers are still thin on ground; local players dominant in chemicals.

Table 3.5. (continued)

Company	Nature of activity and business model	Current products in China	Strategy	Positioning
Degussa	Four out of five divisions; Degussa sees China as the driving economic power in Asia and intends to significantly expand its activities in China in future. Is pursuing a "strategy of three developments": developing employee skills, new technologies and the market	Carbon black, amino acids, polyurethane foam additives, water treatment chemicals, construction chemicals, initiators for plastics production	Planned investment: more than EUR 100 million; Degussa is planning to build a so-called "multi user site" in China that will create the infrastructure for all Degussa's divisions (kick-start function) and also a large number of smaller projects.	Market leadership already achieved in individual divisions
K+S	No salt; sales only of potash, magnesium and fertiva products; not dedicated company, just a representative office working out of Singapore	Nitrogen, potash and magnesium-based fertilizers	Competitors have logistics advantages over K+S in exporting to China. Therefore priority is to expand specialties with bigger contribution margins. No investment planned.	Positioning in standard potash fertilizers is trivial due to fierce competition from Canada and Russia; freight costs too high
DSM	Production and sales through own local company in China	Antibiotics, engineering plastics, resins, caprolactam and vitamins	To expand leading position in vitamins and resins by forming joint ventures	Leadership position in antibiotics and vitamins
KWS	Establishing a branch office to start building a distribution network	Sugar beet seed	To establish licensed production of sugar beet, sunflower and corn seed in China	Leading supplier of sugar beet seed
Fuchs Petrolub	Production and sales, both through own subsidiary and through joint ventures	Vehicle lubricants, industrial and process oils, metalworking/machining lubricants and corrosion protection agents for metal surfaces, greases	To continue to buy out joint venture partners and build up own organization in China	1.2 % share of Chinese lubricants market (3.5 million t)

Table 3.5. (continued)

Company	Nature of activity and business model	Current products in China	Strategy	Positioning
Syngenta	Seed and crop protection segments. Sales of seeds and crop protection products. Production of crop protection agents	Seeds, herbicides, insecticides, fungicides	To increase productivity by offering integrated plant production solutions	Syngenta is the biggest investor in crop protection products in China and operates in more than 25 provinces.

BASF

In 2003, BASF's 3,000 employees in China generated sales of EUR 1.6 billion. This was an increase of 23 percent over 2002. This allowed the group to easily more than compensate for unfavorable exchange rate movements. BASF was amongst the front runners in planning major investments in China. Its big integrated plant in Nanjing will be mechanically completed as early as Q4 2004. With investments totaling US$ 4.3 billion in China (together with partners), BASF heads the ranking of chemical companies investing in China. Its China exposure of 4.8 percent already represents an impressive order of magnitude. BASF, together with its partner Sinopec, is investing US$ 2.9 billion just on its integrated facility in Nanjing, which is similar to the Antwerp site and centers on a naphtha cracker for producing basic chemicals and especially ethylene. Once the plant is on stream, basic chemicals are likely to constitute around 20 percent of BASF's China-generated products portfolio, which will allow BASF to participate in the basic chemicals boom. BASF together with partners is also investing US$ 1 billion in an isocyanate production joint venture (MDI/TDI) and US$ 300 million in a wholly owned polyTHF production at its Caojing plant. In these cases BASF will also profit from the demand for specialty chemicals, especially from the ravenous automotive sector. We therefore predict a sales CAGR between 2003 and 2008 of 17.4 percent for BASF in China.

Bayer

Bayer first moved into the Chinese market back in 1882. Its 2003 sales in Greater China amounted to EUR 1.1 billion. This was in line with the prior year as organic growth and unfavorable exchange rate effects cancelled each other out. Bayer already has a significant exposure to China, the source of 3.9 percent of group sales. 2,100 employees work locally for Bayer in all three of its business segments. The group is concentrating its investments in China on polymers production. Its

China focus in the next few years will increasingly be on the MaterialScience subgroup. The group has a strong position in polymers, a field where local competitors are still thin on the ground. Bayer has a good opportunity to profit from the booming demand for automobiles in China through the local production and subsequent marketing of plastics (specialty chemicals). Bayer has accordingly decided to manufacture its top plastics product Makrolon (polycarbonate) in China starting in 2006. We predict average annual sales growth between 2003 and 2008 of 14.2 percent for Bayer in China.

Degussa

Degussa has been in China since 1933 and has been producing specialty chemical products locally since 1988. The group has 17 subsidiary companies and several production sites. Degussa's more than 1,000-strong workforce generated 2003 turnover in China of EUR 280 million, a 16-percent advance on the prior year (EUR 240 million) despite the negative currency effects. It is strong customer demand that has attracted Degussa to China. The group aims to supply not only the domestic market but also other regions of Asia from its local production. The company sees China as the driving force in Asia, and has targeted a medium-term doubling of the Asian component of its total sales (2003 share: 14 percent). It has defined China as the most important market where it can achieve top-line growth. In May 2004 Degussa signed a cooperation agreement with the Chinese high-tech company Changchun Jida High Performance Materials covering the joint development, production and marketing of high-performance polymers. Degussa is planning to invest over EUR 100 million Euro in the carbon black, pharma amino acids and polyurethane foam business segments. Its local production of carbon black, rubber silanes and polyurethane foam additives is allowing the company to profit from the strong growth of automobile demand in China. Degussa's production of construction chemicals in China is also being boosted by the country's building boom. Degussa is additionally working to transfer the production of commodity fine chemicals products (building blocks) to China, since it is no longer economical to manufacture these in Germany due to the intensification of competition from Chinese rivals. We estimate a sales CAGR between 2003 and 2008 of 9.5 percent for Degussa in China.

DSM

DSM's 3,000-strong workforce generated 2003 sales in China of EUR 273 million. Through its acquisition of Roche's vitamins business, DSM has now put itself on a sure foundation in China. Its China sales are already 4.5 percent of group sales and this proportion will rise due to the full-year consolidation of the vitamins operation in 2004. If we add in the acquired vitamins sales, pro forma sales total EUR 350 million. DSM is expanding its local production in order to strengthen its leadership position in vitamins, antibiotics and resins. Annual caprolactam production is scheduled to increase to as much as 140,000 tons by the end of 2005. Some of DSM's penicillin products already hold a more than 50-percent market share in China. With DSM predicting annual growth rates of 30 to 40 percent for industrial plastics with automotive industry applications, the company is aiming to double its China sales to EUR 600 million by 2008 (estimated CAGR 2003 to 2008: 17.1 percent). DSM is therefore the only company with a formal sales target for China. We are a little more cautious and see a CAGR over this period of 13.5 percent. Its exposure to vitamins and fine chemicals makes DSM one of the companies that are suffering severe price pressure due to the exporting of cheap products from China to Europe.

Fuchs Petrolub

Fuchs Petrolub's 2003 China sales totaled EUR 57.2 million, which represents a double-digit increase over the prior year. The company employs around 300 people in China. Its early commitment to the Chinese market in the late 1980s enabled Fuchs Petrolub to seize key positions as a foreign supplier of lubricants before other competitors began to rush into the market. The company's China exposure of 5.5 percent is the biggest in our coverage. Fuchs Petrolub is planning to invest a sum in the low millions of euros in a plant for manufacturing metal surface lubricants. The company is concentrating its activities in China on supplying lubricants to the metalworking and automotive industries. In the process it is profiting from the rapid expansion of local automobile production. Its focus is the production and marketing of specialties that permit relatively wide margins. With three production sites, a big branch network and third-party distribution, Fuchs Petrolub is superbly positioned in the Chinese market. We accordingly expect a sales CAGR between 2003 and 2008 of 16.0 percent for Fuchs Petrolub in China.

K+S

K+S posted 2003 sales of EUR 36 million in China, mainly from nitrogen fertilizers (fertiva), which was a 16-percent drop from the prior-year total (EUR 43 million). K+S was unable to compensate for the unfavorable exchange rate trend. The proportion of its China sales to total group turnover has shrunk even further to 1.6 percent (from 2.0 percent) due to these currency factors. K+S is betting in future in China on specialty fertilizers. It distributes high-value products in the country such as potassium sulfate, for which it has a market share of around 13 percent. In order to optimize its earnings, K+S prefers to sell standard fertilizers into logistically more favorably located overseas regions (e.g. Latin America). The company only has two sales representatives in the field in China. It is not investing in China and does not consider the country a priority. It follows that the company is unlikely to profit from the China boom.

KWS

KWS had 2003 sales estimated at EUR 2 million in China. At 0.5 percent of group turnover, its China activities represent the lowest exposure of any company in our coverage. Its workforce is probably just one person. The company sells sugar beet, sunflower and corn seed in the region. It is not planning any significant investment in China and maintains just one sales office there. The company has been cooperating with local seed companies on the production and research fronts since the mid-1990s and this has enabled it to become the leading supplier of sugar beet seeds. Its objective is to set up licensed production of sugar beet, sunflower and corn seed in China and to carry out in-house research in the country. The increasing urbanization of China is likely to boost local meat consumption. KWS will be in a position to profit from this trend as it will increase the demand for high-performance varieties to satisfy the expanding need for fodder cereals and other animal feed crops. China is not a priority for the company. We expect KWS Saat's China sales to grow at an average annual rate of 6.8 percent in the period from 2003 through 2008.

Syngenta

Syngenta first got involved in the Chinese market back in 1886. This early commitment has enabled the company to acquire China know-how long before other foreign competitors appeared on the scene. Destocking and currency effects caused Syngenta's 2003 sales in China to fall well below the prior-year total of US$ 100 million. We put the company's China sales at US$ 75 million. This would make its China exposure 1.1 percent of group sales. It has a workforce of 450 in China. The company's biggest seller in the region is its Gramoxone crop protection product, a non-selective herbicide. Syngenta is not planning any significant investment. It has sales and production facilities in China and is active in more than 25 provinces. In order to expand its leadership in the Chinese crop protection market, the company is cooperating on research with local players. Syngenta is aiming to increase its market share in crop protection products and vegetable seeds in order to tap into the strong growth of vegetable exports from China. This industry needs to comply with international standards, which requires the use of high-value crop protection products. Syngenta is already profiting from the fact that its locally produced Gramoxone is now the most widely sold herbicide in China. Even so, China is not necessarily a priority for Syngenta. We predict a sales CAGR between 2003 and 2008 of 7.2 percent for Syngenta in China.

Table 3.6. Data on chemical companies' activities in China in 2003 (Source: Company disclosures, DZ Bank estimates)

Company	Market entry	Production start	2003 sales	As share of total sales	Share of sales locally produced	Work-force
Akzo Nobel	Early 1980s	1989	EUR 510 million; EUR 435 million in PRC and EUR 75 million in Taiwan	3.9%	70%	2,500
BASF	1885	1988	EUR 1.6 billion	4.8%	n. a.	3,000
Bayer	1882 first sales; 1913 first office	1935 (Aspirin)	EUR 1.1 billion; EUR 585 million in PRC, rest in Taiwan and Hong Kong	3.9%	40%e	2,100
Degussa	1933	1988	EUR 280 million	2.5%	20%	> 1,000
DSM	1963	n.a.	EUR 273 million	4.5%	35%	3,000
Fuchs Petrolub	1985	n.a.	EUR 57 million	5.5%	> 67%	300
K+S	n.a.	n.a.	EUR 36 million	1.6%	0%	2
KWS	n.a.	n.a.	EUR 2 million	0.5%	0%	1e
Syngenta	1886	2001	< US$ 100 million	<1.6%	n.a.	450

4 Research and Development in China

Gunter Festel: Festel Capital, Huenenberg, Switzerland;
Harald Pielartzik and Martin S. Vollmer: Bayer MaterialScience AG, Krefeld
resp. Leverkusen, Germany

In the past few years, there has been a sharp rise in foreign investment in research and development (R&D) in China. Chemical and pharmaceutical companies are stepping up their activities in this field. This chapter takes a more detailed look at R&D in China, with special reference to the R&D activities of
foreign companies.

4.1 Scope and Structure of R&D in China

In 2002 R&D spending in China was around RMB 120 billion (some EUR 12 billion) (*Table 4.1*). Spending on R&D thus accounts for 1 percent of gross domestic product (GDP), which is well below the OECD average and extremely low
compared with Europe, the United States and Japan where R&D spending is between 2.5 and 3 percent. Compared with the OECD countries, R&D intensity in
China is particularly low in the high-tech industry. While the pharmaceuticals
sector in OECD countries spends more than 20 percent of value added on R&D,
an OECD survey puts the comparable figure for China at around 2 percent. The
situation is similar in other high-tech industries such as aviation and electronics.

Table 4.1. R&D spending in China 1991-2002

	1991	1995	1996	1997	1998	1999	2000	2001	2002
National R&D spending in China (RMB billion)	15.1	34.9	40.4	50.9	55.1	67.9	89.6	96.0	116.1
Real annual growth rate (%)	-	-	9.5	24.9	10.9	20.3	16.9	9.9	20.5
R&D spending as % of GDP	0.70	0.60	0.60	0.64	0.69	0.83	1.01	1.06	1.10

In China there are 380 research institutes run by the central government and
about 4,000 run by provincial governments and local administrations. The research landscape was completely restructured in 1999 and 2000. This move included the reorganization of many research institutes. Nevertheless, the country's
R&D activities are still essentially state-dominated and shifting to a market-
oriented system is a major task for the Ministry of Science and Technology
(MOST). In 2000, 60 percent of national R&D spending in China was provided
by companies (*Fig. 4.1*). By comparison, in 1995 state R&D institutes accounted

for the bulk of R&D spending and the corporate sector made up only 32 percent of the total. There are more than 1,600 universities in China, 72 of which are run by the central government. The remainder is run by local authorities and various other organizations.

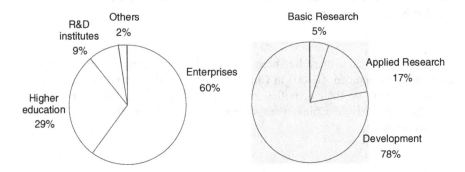

Fig. 4.1. Breakdown of national R&D spending in China in 2000 by institutions and by type (Total: RMB 89.6 million)

By far the largest proportion of R&D spending is specifically for product development (*Fig. 4.1*). Only a small proportion goes into basic research. To promote basic research, the Natural Science Foundation of China (NSFC) was set up in 1986.

The Chinese Academy of Sciences (CAS) is one of China's leading academic institutions and a research and development center for the natural sciences, technological sciences and high-tech innovation. It has a total staff of over 58,000, of whom 39,000 are scientific personnel. CAS was founded in Beijing in 1949 on the basis of the former Academia Sinica (Central Academy of Sciences) and Peiping Academy of Sciences. The CAS comprises five academic divisions (Division of Mathematics and Physics, Division of Chemistry, Division of Biology, Division of Earth Sciences and Division of Technological Sciences), 108 scientific research institutes, over 200 science and technology enterprises and more than 20 support units including one university, one graduate school and five documentation and information centers. All these are distributed throughout the country. 12 branches of CAS were established in Shanghai, Nanjing, Hefei, Changchun, Shenyang, Wuhan, Guangzhou, Chengdu, Kunming, Xi'an, Lanzhou and Xinjiang.

Two of its major research institutes are in Shanghai. The Shanghai Institute of Organic Chemistry (SIOC) was founded in 1950 and covers all aspects of organic chemistry. The Shanghai Institutes for Biological Sciences (SIBS) perform basic research in the biosciences and especially biotechnology. The Institute of Biochemistry and Cell Biology is one of the ten institutes that make up SIBS. It was founded in May 2000 by amalgamating the Shanghai Institute of Biochemistry and the Shanghai Institute of Cell Biology. Its research focuses on polypeptides, proteins and proteomics, nucleic acids, genomics and immunology.

In 1998 CAS announced a 12-year 'Knowledge Innovation Program' to intensify its science and technology innovation capability and overall competitive strength. The program is supported by the Chinese Government and focuses on defining strategic science and technology areas, talent training and contingent building, international collaboration, promotion of technology transfer, development of an innovation culture and construction of adequate science and technology infrastructure. The program includes the clear goal of restructuring the research institutes to create internationally acknowledged high-level institutes. At least three to five will be among the best in the world. By introducing a new personnel management system, CAS has recruited hundreds of excellent young scholars from abroad to implement modern science and technology concepts and to strengthen China's innovative strength. This is directly reflected by the high quality of publications and conference presentations. In the mid to long term, CAS and Chinese universities will become a technology source and exercise a growing impact on industrial R&D. CAS strongly promotes technology transfer and numerous R&D institutions have already been transformed into enterprises within the CAS enterprise system which is supported by internal and external venture capital funds.

Shanghai is the most important R&D location in China. The city's Pudong New Area is expected to become an important hub for the R&D centers of multinational companies in the Asia Pacific region. According to statistics from the Economic and Commerce Bureau of the Pudong New Area Government, more than 70 foreign companies or joint ventures have established their regional R&D centers in Pudong. Most of these are in the high-tech sector, including information technology, biomedicine and new materials research. Statistics indicate that total investment in R&D in the 156 Pudong-based high-tech enterprises hit RMB 2.38 billion in 1998, accounting for nearly 6.4 percent of the enterprises' combined revenue for that year.

The Zhangjiang High-Tech Park in Shanghai was established in July 1992 as a national park dedicated to the development of innovations in high technology. In August 1999 the Shanghai Municipal Government issued "Focus on Zhangjiang", a strategic policy designed to accelerate the park's rate of development. The park's two leading industries are information technology and modern biotechnology and pharmaceuticals, and its principal focus is on developing innovation and entrepreneurship. Its main industries are supported by the national bases located there: the National Shanghai Biotech & Pharmaceutical Industry Base, the National IT Industry Base, the National 863 Information Security Industry Base and the National Technology Innovation Base. By 2005 Zhangjiang will be one of the nation's top high-tech parks, with an innovative and educational atmosphere and state-of-the-art technology. It is also striving to become a world-renowned center for high-tech industries and scholars.

Scientific and technological cooperation between China and Germany has increased considerably in recent years. In October 2000 the German Research Foundation (Deutsche Forschungsgemeinschaft or DFG) and the NSFC set up the "Sino-German Center for the Promotion of Science". This collaboration includes all major German institutions involved in research and the promotion of research (DFG, Max Planck Society, Fraunhofer Society for Applied Research, the Helmholtz Organization of German Research Centers and the affiliated research institutes of the Leibnitz Association and other organizations). Cooperation is coordinated by a joint government commission, which meets every two years, and there are regular meetings of the steering committees for biotechnology, new materials, marine research, environmental technology as well as physical and chemical technologies.

4.2 R&D by Foreign Chemical and Pharmaceutical Companies

In the early to mid-1990s, companies focused on setting up R&D alliances with Chinese organizations,. These days many foreign chemical companies have their own R&D operations in addition to wide-ranging alliances with leading Chinese universities and research institutes. In November 2001 Bayer opened the Shanghai Polymer Research & Development Center in the Jinqiao Export and Processing Zone in Pudong at a cost of EUR 10 million. The company also has research alliances with CAS, Beijing University for Chemical Process Technology, the University of Science and Technology in Hong Kong, the Materia Medica Institute in Beijing and the Kunming Institute for Botany and provided funds to establish a professorship at Tongji University in Shanghai. The specialty chemicals company Degussa has invested EUR 10 million in a 6,900 square meter R&D center in Shanghai, which was opened in April 2004. This is designed to integrate China into Degussa's global R&D network and carry out specific product development work. In addition, Degussa is establishing intensive cooperation arrangements with a range of Chinese companies. One example is the joint development, production and marketing of high-temperature polymers with Changchun Jida High Performance Materials.

A number of other foreign chemical companies have maintained R&D activities in China for many years. In 1997 Novozymes opened an R&D center in the Zhongguancun Science Park in northwest Beijing. Costing EUR 10 million, the facility is involved in the customized development of enzymes and processes for the Chinese market. Toray has an R&D center with more than 100 staff in Nantong which focuses on polymers and fibers. The company is planning to open a second center with 50 employees in Shanghai at the end of 2004. It expects to raise the number of R&D staff in China to a total of 350 by 2006. Dow Corning opened an Application Technology Center with more than 20 employees in the

Songjiang Industry Park in Shanghai at the end of 2002 to focus on the construction, textile, paper processing, oil and gas, personal care, automotive, power generation and electronics sectors. This company also has cooperation agreements on organosilicon chemistry with academic institutes such as Beijing University and the CAS Institute of Chemistry.

General Electric maintains its own R&D activities in China. The China Technology Center (CTC) in the Zhangjiang High-Tech Park is one of General Electric's four centers of research and development worldwide. Occupying an area of 47,000 square meters, the CTC cost US$ 64 million and serves three fundamental roles: advanced research and development of enabling technologies for future GE products and services; engineering support for GE's sourcing activities in China; and education for GE's employees, customers and suppliers. Wacker-Chemie also opened an R&D center in the Zhangjiang High-Tech Park in 2000 and Rhodia has had an R&D center for the Asia Pacific region in the Xinzhuang Industry Park in Shanghai since 2004. Other chemical companies are set to follow suit. DuPont will be opening a 30,000 square meter R&D center for 200 scientists in the Zhangjiang High-Tech Park in 2005 with a total investment of US$ 15 million. This will be its third-largest R&D center outside the United States, after Switzerland and Japan.

BASF does not yet have its own R&D operations in China. However, in 1997 it set up the RMB 30 million "BASF Sino-German Research and Development Fund". So far this has financed more than 40 research and development alliances with Chinese universities and research institutes. The main areas of cooperation are polymers, catalysts, organic pigments and active ingredients for agrochemicals and pharmaceuticals. The fund is administered by a special liaison office in Shanghai.

There are also a number of pharmaceutical companies with their own R&D facilities in China. Roche has opened a new R&D center for 40 chemists in the Zhangjiang High-Tech Park and established R&D alliances with the state-owned genomics centers in Shanghai and Beijing, which conduct research into the genetic predisposition to diseases such as diabetes and Alzheimer's. The new R&D center will support the group's medical chemistry activities and will initially focus on key chemical building blocks for active ingredients and specialized substance libraries. Investing in R&D enables Roche to enter into a dialogue with authorities and opinion leaders in the country, which it expects to boost business in China. The company has some 1,200 employees in China and generates sales of around EUR 100 million with pharmaceutical and diagnostic products. In July last year, GlaxoSmithKline China set up an OTC (over-the-counter) medicines R&D center at Tianjin Smith Kline & French Laboratories, a joint venture funded by Glaxo-SmithKline. Its aim is to excel in creating innovative science-based products to meet consumer needs and support the joint venture's vision of becoming the premier OTC company in China. GlaxoSmithKline has yet to decide whether to establish a prescription medicines R&D center in China.

4.3 Conclusion and Outlook

Previously, multinational companies mainly used China as a base for production plants. But the common practice of undertaking R&D in Europe, the United States or Japan, with only production in China, will change. Given the importance of the Chinese market, an increasing number of chemical and pharmaceutical companies are likely to set up R&D activities in China in the near future. At present, the focus is still on regional R&D service centers to gain a foothold in the Chinese market. The reason is the increasingly fierce competition facing multinationals, which are vying for an advantageous position in the Chinese market. They see China as a country with a great deal of potential and are starting to recognize the importance of R&D as a driving force in this emerging economy.

As a key aspect of their global strategies, local R&D centers are expected to help them respond quickly to market changes due to the proximity to the Chinese market. Using locally based experts to provide individual support and advice enables companies to coordinate technology transfer and speeds up market development because custom-tailored product development means local trends can be taken up quickly. There are many well-trained and highly motivated scientists and engineers and their salaries are low compared with those paid in Western industrialized countries and Japan. Good opportunities for cooperation with universities and research institutes provide access to the latest research findings. Moreover, conditions are favorable for R&D and the government invests billions in universities and R&D centers. Therefore, following in the footsteps of the electronics and software industries, in a few years' time R&D units engaged in more basic research in the chemical and pharmaceuticals industry are likely to be relocated to China.

Nevertheless, investing in China can be problematic because differences in culture and mentality, not to mention the language barrier, mean that setting up alliances or research activities in China requires a good deal of organization. Added to this, there are still some uncertainties surrounding the legal position and the risk of an outflow of knowledge. However, in the past couple of years the Chinese government has made an enormous effort to implement international laws on the protection of commercial rights in China.

5 Chemical Industry Parks in China

Gunter Festel: Festel Capital, Huenenberg, Switzerland;
Yong Geng: Dalian University of Technology, Dalian City, P.R. China

Many investments, especially by foreign companies, are located in chemical industry parks. This chapter characterizes chemical industry parks in China, discusses the main site selection criteria and important aspects for foreign investors, and provides profiles of selected chemical industry parks.

5.1 Industrial Parks and Their Infrastructures

In late 1984, the Chinese government approved the first batch of development zones in 14 coastal cities to provide preferential policies for foreign investors who inject capital into these areas. Since then there has been widespread establishment of larger and smaller industrial zones or high-tech parks. Through industrial parks, firms benefit from economies of scale in terms of land development, construction, common facilities and infrastructure. The industrial zone's importance can be distinguished by its backing, being either a development zone at state level (backed by central government), provincial/municipal level or local/town level. It may even be a privately funded development. Today, according to incomplete statistics from the Ministry of Land and Resources, there are 6,866 development zones across the country (*China Daily,* August 24, 2004).

Industrial parks have played an important role in the national development strategies of many countries and have been irreplaceable where economic development is concerned (*Yang* 2001). In China, such parks have become important showcases and bases for development of an export-oriented economy in the regions. They are the most active areas for foreign investment and serve as bases for parent cities to readjust their industrial structures and renovate old enterprises, as well as providing places where Chinese methods of enterprise management can adapt to the international management standards (*Geng/Côté* 2003).

Generally, an industrial park is land reserved by a municipal authority for industrial development. It usually includes an administrative authority, making provisions for management, enforcing restrictions on tenants and detailed planning with respect to lot sizes, access and facilities. Furthermore, in some countries like China and Thailand, industrial estates have a dual function as production and residential areas. This is different from the North American model in which estates are predominantly manufacturing-based (*Geng/Côté* 2001; *Geng/Côté* 2003). Generally, a typical Chinese industrial estate has an industrial production area, a scientific research area, a residential area, and a business and service area. Some typical characteristics are as follows:

- *Independent:* Industrial estates are separated from their parent cities and equipped with the necessary support infrastructure.
- *Comprehensive:* Industrial estates have been designed for a variety of purposes and for different categories of industry.
- *Superior:* Industrial estates enjoy better infrastructure and investment conditions due to better planning and support.
- *Intensive:* Their activities are typically intensive in terms of capital, revenue and technology.
- *Concentrated:* Industrial estates are concentrated in the eastern coastal areas and in medium to large cities, usually on the periphery of cities and in suburban areas (*Yang* 2001). For instance, there are some 100 such industrial parks in Shanghai Municipality alone. In Zhejiang Province, neighboring Shanghai, authorities have so far approved more than 800 industrial parks, although only 200 are already operational.

For foreign investors, there are certain advantages to investing in chemical parks as opposed to establishing their own chemical sites. These include:

1. *Reduced costs and risks:* Costs are always the biggest concern for foreign investors. With the increasing price of land and construction, it will cost a lot of money for foreign investors to rent a site and build the necessary infrastructure. This is especially obvious in Chinese coastal areas, where the increasing costs of land rental and construction have discouraged foreign investors. However, by investing in existing chemical parks, foreign investors can reduce such expenses and use their funds for other purposes. This is because most of the existing chemical parks have been equipped with the necessary infrastructure to state-of-the-art specifications. In particular, some parks have developed their own emergency response systems which is a key factor allowing investors to reduce their operating risks.
2. *Reduced culture barriers:* Once a company invests in a foreign country, it somehow has to deal with cultural barriers such as language, communication habits and even ways of thinking. If an investor decides to develop his own chemical site, this means finding an appropriate way to overcome these cultural barriers in dealing with local governments, communities and other stakeholders. However, if an investor locates in an existing chemical park, these problems can be avoided by consulting with park management which is usually from the local area and knows the local culture.
3. *Accessible resources:* An existing chemical park usually has good access to the necessary resources (energy and water supply, logistics services, etc) for chemical manufacturing. Park management also has good access to human resources since they usually have a suitable database and know how to best recruit the qualified personnel locally.

4. *Synergies in environmental protection:* When a new industrial park is developed, the Chinese legal system requires park management to prepare an environmental impact assessment report based upon the Chinese standard for official approval. This report has to reflect the potential environmental pressure that will be caused by the park, such as total air and water pollution and solid waste volumes, and describe the necessary waste treatment facilities. This system also asks the park developer to construct environmentally friendly infrastructure, therefore avoiding the potential environmental pollution that might be caused by the tenant companies. If an investor decides to build his own site, he will have to follow this procedure for the whole site, which is both time-consuming and expensive.

Industrial estates currently have one of the highest growth rates in China, with a heavy concentration of both domestic and foreign investment (*Geng/Côté* 2003). They are usually based in areas with an advanced economy, a well-developed industrial base and a good distribution of industrial sectors. For example, while the total area of economic and technological development zones (a major form of Chinese industrial estate) is now 38,604 square kilometers - about 0.04 percent of China's territory - accumulated direct foreign investment in such zones accounts for about 10 percent of the national total foreign capital inflow (*Geng/Côté* 2001). The 1999 per capita productivity of manufacturing businesses in such zones, for example in Shanghai, Tianjing, Dalian, Beijing and Kunshan, was as high as RMB 200,000 (US\$ 1 = RMB 8.2) or more (*Geng/Côté* 2001).

5.2 Site Selection Criteria

One special aspect in China is land legislation which provides two major types of land use rights: the Allocated State-Owned Land Use Right (ALUR) and the Granted State-Owned Land Use Right (GLUR). The ALUR is typically found in the formation of a joint venture with an existing or former state-owned enterprise where the state has "allocated" the right to use the land to its own (state-owned) enterprise virtually free of charge. In return, the state-owned company is expected to develop the land at its own expense, including resettlement of former users such as farmers, development of the necessary infrastructure and payment of a small annual fee to the government (typically in the range of US\$ 0.04 to 0.05 per square foot). In return, the state has the power to withdraw the land use right at its own discretion or increase the annual fee every three years.

In contrast, China's economic reforms that started in the late 1980s saw the introduction of the GLUR as an additional form of land use rights. "Granted" by the state for a one-off payment reflecting the market value of the property for 40 to 70 years (depending on the purpose of the investment), the GLUR currently constitutes the most secure form of land use right as the grantee normally obtains it in the form of developed land directly from the Land Administration Bureau.

Compensation to former users and even some form of infrastructure are mostly included. The grantee has legal protection for the stated term and can even expect compensation in the event of expropriation. However, details about such compensation and the infrastructure included should be carefully checked beforehand. Compensation levels have risen significantly in recent years, as has public awareness of shady land deals, with the subject receiving growing attention even in state-controlled media. As the GLUR is underpinned by a type of market valuation, prices differ very much from location to location. Unit prices can easily reach well above US$ 10 per square foot in attractive coastal high-tech industrial parks, while more remote inland developments offer major discounts on listed land prices, taking them to below US$ 1.80 per square foot.

The site selection criteria for chemical companies and industrial enterprises can generally be divided into operational site factors such as market, raw materials, logistics, real estate, infrastructure, energy and availability of a skilled work force, and functional site factors such as prevailing legislation, investment regulations, taxation, customs and the financial system. A very important factor for any investor is the so-called "US$ 30 million" threshold. While investment projects above this value normally require approval from central government authorities in Beijing, projects below the threshold can be handled by local authorities, normally at a municipal level. The importance of this threshold becomes apparent once a close look is taken at the investment support on local level.

Another important aspect of selecting and establishing a site is to obtain a secure and enforceable right on the underlying property. While China does not have a system of private land ownership and all land is therefore owned by the state, there is an established but still developing system that allows the transfer of land use rights "for value". This system of land use rights, however, requires great attention on the part of the investor as it has undergone significant changes in the past and is still being optimized. Furthermore, the country's fast pace of development has created a huge demand for construction land and, as a result of this, a tremendous rise in land prices.

There are four site selection criteria for chemical parks in China. These include:

1. *Convenient transportation:* The chemical industry provides many raw materials for other industries. Within a chemical park, the material flow (both input and output) is immense. Consequently, convenient transportation is essential. In China, it is usual to build such parks near seaports or river ports with a deep water harbor.
2. *Abundant water resources:* Water resources are the key to developing the chemical industry. This is a typical process industry that requires abundant water for its operations. A chemical park should be able to provide at least 1 to 2 cubic meters per second (*Yang* 2004).
3. *Adequate environmental carrying capacity:* The chemical industry is a large polluter. Although many environmental protection measures have been

adopted by the chemical industry, some pollution is still unavoidable. There-fore, a certain environmental carrying capacity within and around the chemi-cal park is a necessity so that some pollution can be absorbed by the local envi-ronment.

4. *Reasonable distribution:* Currently, the competition between different chemi-cal parks is very fierce in terms of recruitment. Some regions have even devel-oped several chemical parks within their territories. Such a situation results in an ineffective and inefficient use of resources. Nobody can win without a rea-sonable distribution of chemical parks in China.

Chemical park management applies certain criteria in selecting foreign investors, including:

1. *Planning limits:* All the chemical parks have their own master development plans. These define certain limits on tenant companies, such as resource con-sumption. For instance, if a potential tenant consumes more water or energy than can be provided by the park, then he may be rejected.

2. *Product type:* Potential foreign investors should inform the park management about which kind of chemical products they intend to manufacture. If their products cannot be easily integrated into the existing supply chains or are not environmentally friendly, then the investor may be refused.

3. *Environmental performance:* Potential foreign investors should prepare an en-vironmental impact assessment report for their proposed facilities. If this does not meet local government approval, the project will be rejected. Conse-quently, foreign investors have to use state-of-the-art technologies (e.g. ecodesign and cleaner production) to minimize pollution.

5.3 Important Aspects for Foreign Investors

However, over-development of industrial zones has led to financial risks since many of them were built with bank loans. More importantly, many development zones either under construction or in the pipeline are not attractive to investors. It goes without saying that all these developments – although most of them are fo-cused on specific industries – are competing to a large extent with each other for potential investors.

Therefore, a thorough investigation of these parks' true capabilities is highly recommended as brochures and initial information may occasionally be exagger-ated, especially in the case of smaller and inland developments. Not every indus-trial park maintains the same service level. Larger industrial parks are of particu-lar interest for overseas investors as they have established "one-stop agencies" - branch offices of the local authorities that operate under the roof of the industrial park's administration. In this situation, the projects valued below the aforemen-tioned US$ 30 million threshold can normally be approved by these administrations themselves. This virtually guarantees significantly less bureaucracy and quick approval, normally within just a few weeks. The involvement of higher authorities may sometimes result in lengthy procedures lasting as much as several months or even years.

As competition for investment continues among industrial parks and communities, many of them try to impress with a high level of "flexibility" to attract industrialists and meet their requirements. Such flexibility, although normally welcomed and highly appreciated by investors from over-regulated Western countries, should be treated with a great deal of caution and careful evaluation. The Chinese legal system is still undergoing a major turnaround to reflect the country's changed circumstances. It therefore often lags behind the country's actual pace of development and is sometimes even contradictory, leaving room for loopholes and various interpretations. In such an environment, promises are given easily – even in writing. Their validity, however, needs very careful assessment with respect to current and future legislation, technical feasibility and the real authority of the respective official. The latter, in particular, can be very confusing for Western investors and requires long experience in dealing with Chinese authorities. Furthermore, it is typical for China that statements and agreements are made in a rather general way. As the devil is always in the detail, this can be a very frustrating experience for Western managers who want to discuss and agree every possible scenario right from the start. Generality always leaves a face-saving "emergency exit" and room for subsequent further discussion.

It is important to know that contracts and agreements do not have the same rigid meaning and importance in China's business environment as they do in the Western hemisphere. This concept, often regarded by Western investors as some kind of legal insecurity, also has advantages as it applies to both sides. There is always room for discussion and modification. In a fast-developing environment like China with many unpredictable changes in direction, this concept provides an important tool for implementing necessary adjustments whenever they arise.

The above experience becomes even clearer upon first arrival at a typical Chinese industrial park – frequently in an impressive but underutilized administration building at the entrance to or in the heart of the development. The lobby will probably hold a large-scale model of how the industrial park is intended to look in the future. A subsequent site tour usually gives a different picture: vast idle or agriculturally used areas, often even still occupied by plenty of farmhouses. Under such rather visionary circumstances, it is especially important for industrial investors to rely on more than just the will for development and the breathtaking pace of Chinese construction.

Right from the start, the potential investor should clarify and secure which infrastructure is already available or will be available in the short term and, even more importantly, at what cost. A frequent problem is that especially utility and site service infrastructures such as electricity, railway, water supply or wastewater treatment facilities are not scheduled to be built until many years later. Or, if already available at an early stage, they have not been built for a phase-wise development but are instead hugely over-designed and therefore result in uncompetitive charges.

In addition, attention has increasingly focused on environmental issues that have emerged during this period of industrial park construction and operation. Such issues include increased pollution, water treatment costs, safety problems and healthcare costs, the loss of biodiversity and challenges to coastal zone management (*UNEP* 2001). The impact of industrial estates is even graver when coupled with natural resource scarcity issues.

The management of industrial estates and environmental protection bureaus are seeking ways to minimize the impact in the face of worsening environmental problems. These issues have not yet been fully considered and integrated into the planning and implementation process, leading to serious impact and damage both within the estates and to surrounding communities. One solution has been to adopt principles of environmental management, which encourage government and businesses to integrate sustainable practices, comply with environmental regulations and implement a systematic approach to a wiser use of resources (*Geng/Côté* 2003). To implement environmental management, several tools have been developed such as cleaner production (CP), APELL (Awareness and Preparedness for Emergencies at the Local Level) and environmental management systems (EMS) (*Geng/Côté* 2001; *Geng/Côté* 2003). Currently, Chinese industrial estates are exploring these tools as a means to mitigate environmental factors.

Many of the tools have been designed for application at the individual facility level (e.g. cleaner production), but there is increasing recognition that sustainability issues can only be effectively addressed at a multi-industrial and spatial level, for example by eco-industrial development. Some of these concepts, like industrial ecology, are still in their infancy, while other approaches, like EMS, have been in use for the last five years (*Geng/Côté* 2001; *Geng/Côté* 2003).

5.4 Selected Chemical Industry Parks in China

With the increasing demand for chemical products, the Chinese government has recognized the significance of developing chemical parks so that the chemical industry can share infrastructure and reduce costs. According to a recent survey, there are at least thirty chemical parks in China and over 20 under construction (*Yang* 2004). Most of them locate on China's eastern seaboard, which reflects the fact that the economy in east China is much stronger. Such chemical parks are categorized in four types in terms of their features, as detailed below:

- *Type 1*: Within this type of chemical park, the anchor industries are refining and ethylene companies, usually large-scale petrochemical groups. With a comprehensive plan and systematic development, various companies from different chemical industry sectors locate in one area on the basis of the supply chains. Typical parks include Yangzijiang International Chemical Industry Park (Zhangjiagang, Jiangsu province), Lingang Chemical Industry Park (Cangzhou, Hebei province) and Qilu Chemical Industry Park (Shangdong province).

- *Type 2*: Such parks usually locate in areas with abundant water resources and deep water harbors. Most of this type are developed in the coastal regions or along large rivers, such as the Yangzi. With convenient geographical conditions and a strong competitive ability, such parks have attracted some multinational companies. Typical parks include Shanghai Chemical Industry Park, Nanjing Chemical Industry Park and Taixing Chemical Industry Park (Jiangsu province).

- *Type 3*: These parks are based on former state-owned chemical companies. They usually focus on manufacturing specialty chemical products, which is attractive to some investors. However, the production facilities are generally old and need to be upgraded, although their infrastructure could still meet investors' demands. Typical parks include Fushun Petrochemical Industry Park (Liaoning province) and Maoming Petrochemical Industry Park (Guangdong province).

- *Type 4*: This type includes those newly developed chemical parks with a relocation of old chemical industries. Most tenants within such parks were originally located in the downtown area of a city. They were major polluters and their facilities needed to be upgraded. In order to improve the urban environment, the municipal government asked them to relocate to the newly developed chemical parks where an environmentally friendly infrastructure has been designed and developed. The financial aspect of such a relocation is that such companies can acquire funds by selling their downtown land (usually at a higher price). They can then use these funds to rent new space within a park (usually at a lower price), retrofit their facilities and install new production technologies. Typical parks include Changzhou Chemical Industry Park (Jiangsu province), Shangyu Chemical Park (Zhejiang province) and Xinghuo Chemical Industry Park (Jiujiang, Jiangxi province).

Some chemical industry parks (Shanghai Chemical Industry Park, Xinzhuang Industrial Zone, Nanjing Chemical Industry Park and China Fine Chemical Industry Taixing Park) are described in detail below.

5.4.1 Shanghai Chemical Industry Park

The Shanghai Chemical Industry Park (SCIP) was the first industrial zone specialized in the development of petrochemical and fine chemistry businesses. It is also one of the four industrial production bases in Shanghai. Located on the north side of Hangzhou Bay, it has a total planning area of 30 square kilometers and is built with the advanced development concept of a world-class and large-scale chemical park. It aims to provide investors with an optimum investment environment by combining production projects, public utilities, logistics, environ-

ronment by combining production projects, public utilities, logistics, environmental protection and administrative services. SCIP has set out to be one of the largest, best integrated and most advanced world-scale petrochemical bases in Asia (see *Table 5.1*).

SCIP is one of the industrial projects with the highest investment levels during the 10th five-year plan period. The aggregate capital investment for the phase 1 program is RMB 150 billion. The industrial output of SCIP will reach RMB 100 billion once all approved projects have gone into operation. To date, BP, BASF, Bayer, Huntsman, Air Products and Chemicals, Ltd., SUEZ, Vopak, AIR LIQUIDE, Praxair and other world-renowned multinational petrochemical and utilities corporations have started projects in SCIP with an aggregate investment of over US\$ 8 billion.

Table 5.1. Petrochemical and chemical companies at SCIP (as at May 31, 2004)

Companies	Investment (US$ million)
Shanghai Secco Petrochemical	2,704
Bayer Coating System (Shanghai)	142
Bayer MaterialScience (Shanghai)	564
Bayer Polyurethanes (Shanghai)	1,290
Shanghai Lianheng Isocyanate Co. Ltd. Shanghai BASF Polyurethane Co. Ltd. Huntsman Polyurethanes Shanghai Ltd.	1,120
BASF Chemicals Company Ltd.	304
Shanghai Tiangyuan (Group) Huasheng Chemical	445
Lucite International (China) Chemical	110
Shanghai Tianyuan Chemical Factory	30
Total	**6,709**

The SCPI will also be a major R&D location. For example, the Sinopec Shanghai Petrochemical Institute will build the National Engineering Research Center for Basic Organic Materials in SCIP.

5.4.2 Nanjing Chemical Industry Park

The government of Nanjing in China's Jiangsu province is investing around US\$ 130 million over the next three years to build infrastructure and prepare land for phase one of a new chemical industry park. This is located on the Yangzi River next to the US\$ 2.9 billion petrochemical complex now being built by BASF and Yangzi Petrochemical. Initially, an area of close to 4 square miles will be readied. Later, officials envisage that it will grow to more than 11 square miles. The product focus is on petrochemicals, basic organic chemical raw materials, fine chemicals, polymers and new materials. Investors are Celanese (acetic acid plant), DSM / Nanjing Chemical Industrial (caprolactam unit), BOC / Nanjing Chemical Industrial (carbon monoxide), Shaw Group (pipe fabrication facility) and Red Sun (pyridine, paraquat, diquat and chlorpyrifos). The park will compete for investment with the Shanghai Chemical Industry Park, located less than 150 miles away in Shanghai's Caojing district.

5.4.3 China Fine Chemical Industry Taixing Park

Taixing is situated in the middle of Jiangsu Province, 180 kilometers away from the port of Shanghai providing access to the East China Sea. Located 8 kilometers west from downtown Taixing, the China Fine Chemical Industry Taixing Park was set up in 1991 on the banks of the Yangzi River. The current planned area of the industrial park is 20 square kilometers, with a long-range plan for 38 square kilometers. 65 enterprises are located in the park; two thirds of them are wholly foreign-owned or joint ventures. The emphasis is on chlorine and alkali, dyestuffs, pigments, pharmaceuticals, intermediates for pesticides, water treatment agents, electronic chemicals and oil chemicals. There are many large enterprises nearby, such as Yangtze Petrochemicals, Gaoqiao Petrochemicals, Jinshan Petrochemicals and Yizhen Chemical Fibers.

6 Trends in the Chinese Fine Chemicals Market – Opportunities and Threats for the European Fine Chemicals Industry

Dahai Yu: Degussa AG, Trostberg, Germany

Recent developments in the Chinese fine chemicals market and its impact on the global market are reviewed. An attempt is made to forecast the future development of this industry in China. Some recommendations for the European fine chemicals industry are provided.

6.1 General Introduction to the Fine Chemicals Market

Since the early 1990s the global chemical industry has been restructured into two major segments: commodity chemicals and fine and specialty chemicals. Commodity chemicals are sold strictly based on their chemical composition and specifications. Specialty chemicals can be single molecular compositions or complex blends and are often sold based on their performance. In addition to product sales, specialty chemical companies also provide innovative services and integrated solutions to their customers. Fine chemicals, positioned between commodity chemicals and true specialties, are high value-added products, ranging in price from a few dollars per kilogram (basic intermediates) to thousands of dollars per kilogram for active pharmaceutical intermediates (APIs). They are usually produced in multipurpose plants. For pharmaceutical and food applications the plants may require special certifications such as those awarded from FDA inspections or comply with current Good Manufacturing Practices (c-GMP) guidelines. Like commodity chemicals they are usually sold based on their specifications. However, it is also essential for the business to provide additional services like R&D and to ensure the speed of process development. Therefore, a broad technology platform and a good image are required.

The global fine chemicals market (including captive and merchant market) was estimated to be worth about US$ 70 billion in 2004 and is expected to grow at an average rate of about 5 percent through 2008 (*Fig. 6.1*). Fine chemicals producers serve the pharmaceuticals, agrochemicals, plastics and other industries. While the United States represents the biggest fine chemicals market, the leading producers (Degussa, Lonza, DSM etc.) are based in Europe due to historic reasons. The merchant fine chemicals market is still a highly fragmented one. The top ten producers have a combined market share of less than 20 percent.

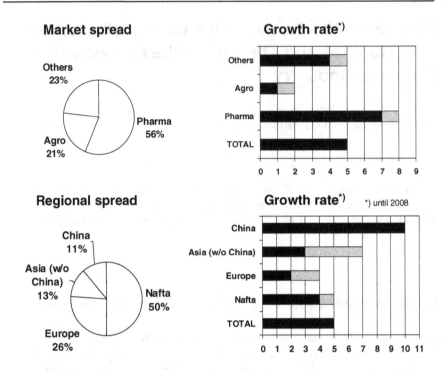

Fig. 6.1. The fine chemicals market in 2004 (in total about US$ 70 billion)

Since the late 1990s there have been numerous M&A activities within the fine chemicals industry. These activities were driven by the perception that the fine chemicals market is an attractive market with quite significant growth potential. The above-average growth was a result of a refocus of the pharmaceuticals and agrochemicals industries on their core competencies. The magic word at that time was "out-sourcing". The following are considered the most important key success factors for the fine chemicals business:

- customer intimacy
- technology tool box
- reputation and size
- development and scale-up capability
- service-oriented approach.

The fine chemicals market is currently facing dramatic challenges. The reduced outsourcing trend of pharmaceutical companies and capacity expansion have created significant overcapacity. In addition, the Asian producers have grown faster than expected. The unfavorable euro / dollar exchange rate has further enhanced the competitiveness of Asian producers. Therefore, restructuring became necessary in Europe, including divestments, plant shut-downs and write-offs of goodwill.

Today the fine chemicals market in general still offers attractive margins. It is important to emphasize that the variable costs (raw materials and utilities) for Chinese players are often more or less comparable to those of the Western producers. The European players suffer from very high fixed costs as the plants' operation rates are currently low and wages are significantly higher than in China. Many companies still maintain a significant marketing staff and R&D capacity as additional services to defend their market position. These cost positions are justified by relatively higher margins. However the question is: Will the market still honor this service in the future? Chinese competitors usually do not have to bear these costs, or the relevant costs are significantly lower compared with their European peers.

6.2 The Chinese Fine Chemicals Industry and Recent Developments

Historically there were no specifically defined segments for the chemicals market in China. Most chemical companies were organized under the roof of the Ministry of the Chemical Industry, and some others were either state- or commune-owned. These companies fulfilled dedicated tasks within the state-planning program to produce the relevant materials. With the fast development of a national economy, reform of the total chemical industry is needed. On the petrochemical side, two giants - Sinopec and PetroChina - have been established. These should have the critical mass to face both national and global competition. The fast development of manufacturing industries for textiles, electronics and shoes has also generated a huge demand for specialty and fine chemicals. China still imports a lot of fine chemicals due to a lack of technologies and know-how. Therefore, focused development of a fine chemicals industry has been put on the agenda in recent years.

The Chinese fine chemicals industry has an estimated output worth about US$ 12 billion.[1] This represents a significant part of the world's fine chemicals industry and cannot be ignored. There are as many as 20,000 producers in China capable of manufacturing fine chemicals. Even though most of them have very poor technological competence and production capabilities by Western standards, more than 500 have GMP-certified production plants. Most of them concentrate on producing basic intermediates and active ingredients for pharmaceutical and agrochemical industries in China.

Numerous investments have been made and new companies have been established and built up in the last five years. The average company has sales of less than US$ 10 million and under 1,000 employees. Some companies have a turnover exceeding US$ 50 million with focus on the more attractive European and U.S. markets.

---

[1] The definition of fine chemical products in China differs from that in Europe. Some products would be classified as specialty chemicals according to the European definition.

A high growth rate is one obvious characteristic of the Chinese fine chemicals industry. The industry has been growing at an annual rate of more than 10 percent over the past ten years, and it is most likely that this growth rate could be maintained for the next five years. The reasons are China's generally strong economic growth, production relocation of Western chemical companies and the increasing trend among downstream manufacturers like the pharmaceuticals and agrochemicals industries to source from China.

The low labor costs compensate in many cases the cost advantages derived from the advanced technologies of Western companies. However, low investment, low-cost and often low-tech production also result in very low entry barriers for other new emerging competitors in China. As the market information system has not yet been fully developed in China, this can lead to problems. Very often, as soon as the margin for one specific product becomes attractive enough, too much new investment is initiated. This can lead to a dramatic overcapacity and fierce price competition for this specific product.

Many Chinese companies manage to have quite attractive margins. It is not rare to find a company with EBITDA margins higher than 20 percent, but the companies often have a very weak cash position as the investments have been financed to a large extent by bank loans with high financing costs. In the event of possible price erosion or other market turbulence, a company will face real financial problems. This is the other weakness in addition to the low technology and low entry barrier characteristics in China.

Chinese producers still suffer from the lack of a track record. Most companies have a relatively short history and are unknown to Western companies. The different business and cultural practices in China are additional barriers to building business relationships. The reliability of supply is still low but has been improving over recent years with the help of potential customers (customer audit, certification and technology transfer etc.).

Chinese fine chemical producers have made a lot of effort to increase their competitiveness in recent years and have closed the development gap to a large extent. The following major trends can be summarized.

Improving Marketing and Sales Competence

Historically, marketing and sales functions did not play an important role in Chinese companies. The businesses were often based on the opportunistic "price selling" approach. Marketing and sales skills were very poor. The Chinese local fine chemicals market is extremely price-driven and this situation will unfortunately remain in the future.

On the export market, Chinese producers are increasingly accepted for supplying standard intermediates, but they are still not yet able to gain access to the more attractive high-margin business for advanced intermediates and active ingredients. Cultural differences and language barriers also make business development difficult. Chinese producers recognized very quickly that gaining marketing expertise is a key to business success. They began not only to rely on the experience of Chinese and European trading companies but started with their own direct marketing and sales activities. By doing so, they were not only able to gain a lot of market information and increasing acceptance by foreign customers but also improved their marketing and sales competencies. Some Chinese companies even hired local European managers as native speakers in order to develop their business quickly.

Boosting the Exchange of Information

The Chinese government supports the growth of the fine chemicals industry actively, not only by establishing new priority investment zones but also by boosting the exchange of information in the country. Some institutions, originating from governmental and ministerial organizations (e.g. China Chemical Information Centre), use their knowledge and focus now on developing businesses based on technology and market information. Some well-known newspapers and trade magazines (similar to Chemical Marketing Reporter or Chemical Week) are newly published and commonly accepted in China. Internet-based research media (e.g. www.finechem.com) have also been established. These organizations perform dedicated desk and field researches for single clients, both Chinese and foreign companies. These initiatives are well accepted by Chinese producers for gathering local information, but they still lack information and intelligence about markets outside China. This gap will be closed as Western information companies (e.g. SRI, Chem-Systems etc.) start to cooperate with their Chinese peers for mutual benefit.

Another possibility for obtaining market information is through attendance of the traditional exhibitions (like CPhI or Informex). If one carefully follows the events over the past ten years, one can clearly see that the number of Chinese exhibitors has steadily increased. Ten years ago, you might have found some unknown Chinese companies (mostly Chinese state-owned national trading companies) sitting in corners with samples. Now almost one quarter of the exhibitors comes from China. They have professional booths of comparable size to their European counterparts. A similar trend can also be observed at exhibitions in Asia, where the regional network has been developed quite well.

Implementing an IT Platform

The quick development of IT technology has probably helped the Chinese fine chemicals industry more than it has affected the European industry. An increasing number of Chinese players use this media to develop business. They begin by creating their own home pages, sending e-mails for sales promotion and end by conducting professional Internet-market research. European industries even help this development as they have established e-bidding processes as a corporate purchasing tool and also use these Internet-based media for China sourcing.

Upgrading Technology Competence

Research and development competence is one of the key differentiating factors for global fine chemicals companies. To a certain extent, Singapore serves as a role model for China. In addition to systematically creating industry waves in electronics and chemicals, the Singaporean government more recently pushed the development of R&D hubs and switched to knowledge-based pharmaceuticals and biotechnology industries. This approach has been successful in attracting foreign investment and creating new jobs. Chinese producers have recognized the importance of this competence for their future business development. Increasing numbers of Chinese companies have begun to build R&D labs and hire development chemists, even with PhD degrees.

Two factors have eased this development. China has a very good university and research institution landscape. A lot of Chinese students with European and U.S. educational backgrounds are increasingly returning for better jobs and business opportunities, and bringing back to China state-of-the-art technology and the highest level of scientific knowledge. This is a result of open door policies initiated by Deng Xiao Peng in the late 1970s.

Defending Intellectual Property

While it is understandable that Western companies are concerned about intellectual property (IP) protection in China, we might have a different perspective if we put this aspect together with the development stage in China. IP issues were also a concern in the early stage of industrial development in Europe and the United States. Significant improvements have been reached, particularly in the last 30 years. With increasing development capabilities in China, more attention has been given to protecting intellectual property. This is a normal development. The number of IP law suits in China has increased significantly. Some recent successful Chinese law suits (e.g. Syngenta sued two Chinese companies for IP violation) boosted Western confidence. But China still has a long way to go. I personally view the further development of IP protection as the dominant factor for the development of the Chinese fine chemicals industry.

Improving Environmental Compliance

Environmental standards are very strict for foreign investments, which are favorable for attracting new technologies to China. Local companies are also forced to comply with these standards and renovate accordingly. If not, they will be forced out. The problem has been the stringency of enforcement. While central government is determined to implement the guidelines, provincial and local governments still play a different role and try to help local industries. Recent directives with the aim of shutting down the small carbide plants and improving the environmental standards of the bigger units are a positive example showing how central government tries to avoid wasting resources and regulate the investments of carbide plants induced by growth in PVC demand. The pollution problem in China needs to be solved. This will not only have a positive impact on environmental aspects, but will also help to consolidate the market and achieve healthy economic growth.

6.3 Scenarios for Future Developments

There could be different scenarios for the fine chemicals industry in the next five to ten years. The uncertainties for future development depend on the general economic situation, the future of the political system in China and improvement of the financial system, which will also affect other industries. However, the fine chemicals industry overall should be less affected by the macroeconomic directives from central government as the industry is extremely diversified and fragmented.

It is most likely that the Chinese fine chemicals industry will maintain its high growth rate and become more export-driven. The export market will remain attractive for Chinese producers. Moreover, a major part of Western intermediates production will quite surely be relocated to China as a result of competitive market development. The fact that most Western companies are following this trend and will hence force investment in China will also accelerate west-east migration and increase market growth in China.

Like their European and U.S. peers, Chinese fine chemicals producers will start evolving into global players. A few companies in China with sufficient financial and management resources will invest outside China and even acquire foreign companies to gain market access.

Fierce price competition will continue in China. This will be true particularly for segments with mature products, low technology and low entry barriers. This will be harmful to the Chinese economy as it wastes resources and creates unnecessary pollution. In addition, this will also have an extremely negative impact on the global market. After China entered the WTO, trade barriers were reduced, and more and more Chinese products could be exported easily. This unhealthy competition can only be reduced if the Chinese government takes macroeconomic and regulatory action. A more restrictive finance policy will also help to direct investments in the right way.

Raw materials, crude oil and electricity will influence the future competitiveness of the Chinese fine chemicals industry much more than that of its European peers. China will remain an importer of crude oil and commodity chemicals. However, increasing production efficiency and full integration of the local petrochemical base will help the fine chemicals industry reduce raw material costs.

It is commonly believed that the Chinese currency (Renminbi) is undervalued. A revaluation would most probably reduce the attractiveness of China as a production base. As most China-related business is carried out in U.S. dollars, a stronger dollar coming from its current historic low level will also help the European industry.

Leading-edge technologies will be developed and new competitive producers with increasingly good access to the global market will emerge. The technology gap between Chinese and Western companies in the high-tech field will be much less than expected. Many research areas like material science, polymer science and biotechnology are already covered by the Chinese research community, and start-up firms have been derived from them. Many scientists leverage their know-how with chemical companies and jointly develop new businesses. On the whole, the Chinese research community has become more commercially driven.

The Chinese fine chemicals industry will be fully integrated and become an important part of the value chain in the global fine chemicals market. It is estimated that the output value of the Chinese fine chemicals industry will grow from about 15 percent today to more than 25 percent of the global market by 2010.

6.4 Opportunities and Threats to the European Fine Chemicals Industry

The Chinese fine chemicals market with new emerging producers offers a historically unique opportunity to reshape the global fine chemicals industry. New technology (like the digital mobile phone or biotechnology) is usually a vehicle for fundamental change in the industry. One can also use an underdeveloped region as a starting point for strategic moves to change one's global market position, which otherwise is difficult to accomplish.

An early mover can use its technology to take the lead position, develop the market and accelerate the necessary consolidation. This can be done by making an investment or forming a joint venture with a Chinese producer, an approach which has been successfully practiced by companies in other industries (Volkswagen, Siemens etc.). In China, Volkswagen holds the highest market share in a single country and now produces more cars there than in Germany. This approach seems to be attractive but also bears a lot of risks, especially for the fine chemicals companies. Industry consolidation in China with a huge number of small commune-based firms could be very painful and time-consuming. Very low fixed costs and some other soft factors could prohibit plant shut-downs. As a result, fierce price competition is expected before consolidation. Hence, pay-back of an investment in China is not guaranteed. In order to sustain competition, a minimal cost position is essential in China.

Forming a joint venture with a leading Chinese manufacturer is a viable option. Although there is a growing view that Western companies should choose a wholly owned foreign entity (WOFE), the joint venture still offers some advantages as the Chinese partner has local market knowledge and adequate political liaison at local and provincial level. This could help to reduce the pay-back time. It is crucial to create a common long-term view about the business right at the beginning and then to pursue this. It is not always true, as often stated, that Chinese partners are less reliable. Foreign companies often change their strategy and personnel which is also not helpful for relationship building in China. Relationship building is still very important in China and it takes time. The limited scope of joint ventures is sometimes an obstacle to reaching a jointly shared view of future business development.

Despite IP concerns, it is essential for the European producer to develop new technologies in China. China offers a large number of well-educated scientists. While managers with a Western mindset are relatively rare in China and highly appreciated by global companies, the Chinese scientific communities are still underestimated. Only a few chemical producers commit themselves to actively participating in research and development there. BASF started a cooperation program several years ago, and Degussa inaugurated its R&D centre in Shanghai in 2004 to foster the development.

Certainly, a large proportion of European-based business portfolios is already very mature. With growing Chinese competition some production will be forced to shut down. In this case, China offers a good opportunity to relocate production. This relocation only makes sense, however, if one owns the leading technology and can achieve a clear cost position under Chinese conditions. The challenge is also to manage the relocation in a socially responsible way as it will result in job cuts in Europe.

There are plenty of opportunities and risks linked with the development of the Chinese fine chemicals market. Each company needs to conduct its own portfolio analysis and develop its own specific footprint for its China strategy, as the Chinese market is still unknown to most of Europe. Also, there are often concerns among managers regarding the uncertainties and higher risks associated with Chinese activities. A successful China strategy must be supported by top management's commitment and the strong involvement of senior management.

When I share my views with executives of European chemical companies, I get the impression that some of them are still unsure about whether they should rethink their strategies and put more focus on developing their Chinese activities. The point I want to make here is: European players do not really have the luxury to choose whether they should go to China or not. They cannot neglect the Chinese market if they want to continue playing a leading role in the future.

On the other hand, China does not need a single individual Western company (particularly with the current China hype) to develop a specific technology as most of the technologies already exist in China.

6.5 Conclusion

The Chinese fine chemicals industry is still in an emerging stage and keeps on growing at high rate while changing the industry structure at the same time. The Chinese producers learn fast and reduce the gap to the western producers. As the development of the Chinese fine chemicals market determines the future of the global fine chemicals market, it is essential for western fine chemicals producers to be engaged in China. A dedicated strategy focusing on the own competence and a detailed market knowledge help to reduce the risk and to ensure the success of such a venture.

7 Chemicals for China's Chip Industry

Klaus Griesar, Merck Electronic Chemicals Holding GmbH, Darmstadt, Germany

China's semiconductor market is expected to reach US$ 90 billion in 2008 with a compound annual growth rate of more than 20 percent between 2003 and 2008. This growth will create increased demand for chemicals used in the production of semiconductors and offers great opportunities for chemical companies as suppliers of electronic chemicals.

7.1 The Semiconductor Industry – Market Growth

The remarkable characteristics of transistors that fueled the rapid growth of information technology is that their speed increases and their cost decreases as their size is reduced. In 1965, Gordon Moore predicted that circuit density or the capacity of semiconductors would double every eighteen months (*Moore* 1997). Moore's Law has held for more than 35 years. Semiconductor technology created new market opportunities and new applications within existing markets. New industries appeared, such as personal computers, which would not have been economically viable using other technologies. The microchip industry's sales growth has averaged 17 percent annually over its 40-year history. *Fig. 7.1* shows the drivers of the integrated circuit (IC) market since 1970.

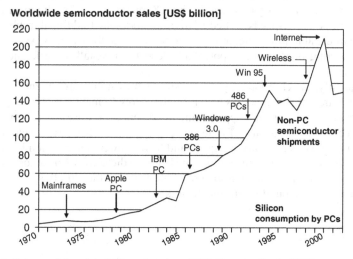

Fig. 7.1. Drivers of the global IC market since 1970 (Source: *Cooke* 2003)

7.2 Technological Innovation in the Microchip Industry

The semiconductor industry is characterized by constant technological innovation and improvement. In general, chipmakers have focused on three areas of technological improvement: smaller linewidths, more reliable interconnects and larger wafer sizes.

7.2.1 Smaller Linewidths, Smaller Chips

Faster-than-expected advances in lithographic technology have accelerated the rate at which chipmakers can shrink device sizes. The economic benefits of this trend are enormous as the ability to cram more chips onto a given space of silicon greatly decreases costs per unit. For example, moving from production of devices with 0.18 micron linewidths to 0.13 micron linewidths effectively yields twice as many chips per wafer. The industry aims for additional gains in yield in the current reduction from to 0.13 micron and 0.09 micron linewidths.

7.2.2 More Reliable Interconnects: Copper Technology and Low-k Dielectrics

Semiconductor interconnect technology has seen recent advances as well. For most of the semiconductor industry's short life, circuit lines that connect transistors and other chip components have been formed with aluminum metal. These thin aluminum lines are protected from each other with an insulating material, usually silicon dioxide. This basic circuit structure worked well through many generations of computer chip advances. As aluminum circuit lines followed the Moore's Law curve and began to approach 0.18 microns in width, the main barriers to higher operating speeds shifted from transistors - the traditional trouble spot - to delays in the aluminum interconnect and increases in the capacity between adjacent conducting lines that can potentially lead to cross-talk (interference) between the lines.

Consequently, the chip industry is now slowly moving toward an entirely new interconnect strategy that employs new insulating (low-k) materials and copper interconnects: While the use of copper as the interconnect promises to improve chip performance mainly by means of reduced resistance, new dielectric materials between the metal wires can improve speed and lower power consumption by reducing capacity and improving insulation. Such low-k dielectrics (insulators with a low dielectric constant k) prevent the cross-talk between the closely spaced wires in the smaller generation of electronic devices.

7.2.3 Bigger Wafers, Higher Yield

Over the years, chipmakers have aggressively increased the size of silicon wafers as a means of efficiently ramping up production volumes. In the early 1970s, the average wafer was one-and-a-half to two inches in diameter. Today, most wafers are either six or eight inches – 150 or 200 millimeters, respectively. The industry is currently building the first generation of fabs using tools to handle 12-inch (300 millimeter) wafers. The nascent transition will greatly enhance productivity and efficiency, as the larger 12-inch wafers will yield approximately twice as many chips as their eight-inch counterparts.

7.3 Electronics Chemicals and Semiconductor Manufacturing

7.3.1 The Electronics Industry

The interrelationship of chemicals, materials, components and equipment is sometimes confusing. *Fig. 7.2* displays an overview of the major segments of the electronic equipment or hardware side of the electronics industry.

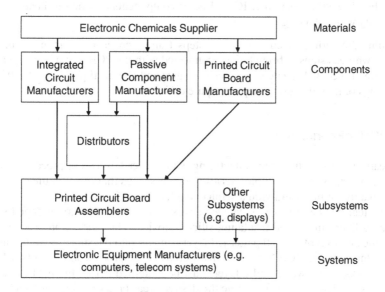

Fig. 7.2. Value chain of the electronic equipment industry

Within this industry, vertical integration is a complex and changing issue. Traditionally, the large electronic equipment manufacturers (called original equipment manufacturers or OEMs) supplied much of their own needs in printed circuit board (PCB) assembly and fabrication through in-house or captive operations. The largest, most notably IBM, also manufactured their own integrated circuits (ICs). On the other hand, each of the boxes in *Fig. 7.2* also represents a strong set of independent or merchant suppliers. There is a general trend away from vertical integration and toward the OEM acting as coordinator of multiple tiers of independent subcontractors.

Specialty as well as commodity chemicals are used in virtually every step of the manufacture of integrated circuits and printed circuit boards. The general manufacturing steps for the production of integrated circuits (ICs) and printed circuit boards (PCBs) are:

(1) Wafer Manufacturing: Manufacture of Si or other substrate and cut into wafers
(2) Integrated Circuit Manufacturing: Series of processes to deposit conductors, semiconductors and insulators to form transistors and wiring
(3) IC Packaging: Chips encapsulated for protection, leads attached, multichip modules formed
(4) Printed Circuit Board Manufacturing: Substrate manufactured, components interconnected, wiring produced
(5) Final Device Assembly: IC and other components attached to board, usually with soldering

According the outline given in *Fig. 7.2*, steps 1 and 2 cover the IC manufacturing process, whereas steps 3 through 5 are connected with the PCB manufacturing and assembling process. The chemicals, technologies and chemical producers for these two groups of manufacturing steps are quite different.

7.3.2 Wafer Fabrication

The heart of semiconductor manufacturing is the wafer fabrication process, where the wafer is exposed to a complex series of process steps that result in the simultaneous creation of many individual semiconductors.

The integrated circuit may be viewed as a set of patterned layers of doped silicon, polysilicon, metal and insulating silicon dioxide. The fabrication process involves the creation of 8 to 20 patterned layers on and into the substrate, ultimately forming the complete IC. In general, many steps in the process should only affect specific areas of the wafer and a layer must be patterned before the next layer of material is applied on chip. To define the desired areas on the wafer (and block the remaining areas), a series of masking layers are used. The process used to transfer a pattern to a layer on the chip is called lithography.

The process of photolithography involves the use of a material called a photoresist to generate a specific pattern on the surface of the wafer. A photoresist is a light-sensitive material which can be processed into a specific pattern after being exposed to light energy in the shape of the desired pattern. Once the photoresist has been patterned, the wafer will be covered by photoresist only in specific areas while the remainder of the wafer is uncovered. Subsequent process steps will affect only the uncovered areas where there is no photoresist. The process allows the transfer of shapes created using CAD software onto the surface of the wafer with submicron resolution. Photolithography is useful because it can transfer the pattern to the wafer surface very quickly. To illustrate the fabrication steps involved in patterning a wafer surface through optical lithography, let us examine the process flow shown in *Fig. 7.3.*

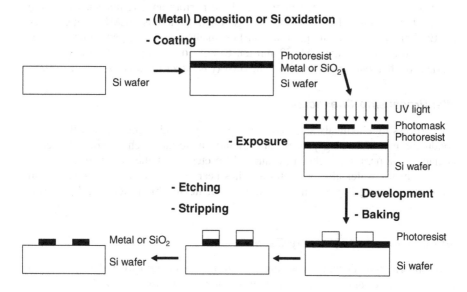

Fig. 7.3: Process flow for patterning a wafer through optical lithography

Since each layer has its own distinct patterning requirements, the lithographic sequence (deposition, coating, exposure, development, baking, etching, stripping) must be repeated for each layer, using a different mask. The recently developed process step known as chemical mechanical planarization (CMP) uses a polishing procedure involving slurries of chemical compounds to smooth the surface of a wafer after each metal interconnect layer is created. As linewidth geometries have shrunk, CMP has grown in importance.

Materials preparation and wafer processes constitute the "front-end" of the chip manufacturing process. The "back-end" of the process begins when the finished wafer is mounted onto a carrier and cut up into individual devices, or dies. "Back-end" processes include packaging and functional testing.

7.3.3 Selected Market Segments of Electronic Chemicals

Photoresists

The primary challenge in the design of photoresists is to find materials which simultaneously possess the following four properties: good optical transparency, good photosensitivity for latent image formation, suitable solubility in the stripping process and good etch resistance. For the materials designer, each new lithographic wavelength has demanded and will demand the development and optimization of – often completely - new materials. The optical properties and behavior of materials change with wavelength, and the performance requirements for the different applications become more stringent and numerous. SEMI (Semiconductor Equipment & Materials International) estimates that a new generation of photoresist costs an average US$ 32 million to develop, compared to a photoresist market worth about US$ 640 million in 2001 (*Chemical Engineering News* 2002).

Wet Processing Chemicals

Wet processing chemicals are used in the etching and cleaning steps of the semiconductor manufacturing process and include materials such as solvents acids, etchants, strippers, and related products. The etching of the circuit pattern on a semiconductor wafer after a photoresist has been applied is an important step in the manufacturing process that is critical in achieving high yields and performance.

Atmospheric and Specialty Gases

Atmospheric and specialty gases are used in the semiconductor industry across a broad number of applications. The largest volumes of gases are used for the protection of wafers from atmospheric exposure during the manufacturing process and to carry or dilute other gases. Nitrogen is used most in terms of volume, followed by hydrogen, oxygen, helium, and argon. The gases used in semiconductor manufacturing are not categorized as specialty gases in the traditional sense, but these gases require high levels of purity, thus rendering them highly specialized.

Other gases are used in the semiconductor manufacturing process as dopants, dry etchants, and in chemical vapor deposition (CVD).

Materials for Chemical Mechanical Planarization (CMP)

More recently, polishing in the form of chemical mechanical planarization (CMP) has become more important, as linewidths on integrated circuits have decreased below 0.25 microns and have begun to contain multilevel interconnects. Wafer polishing includes both a chemical and mechanical action on the surface. The goal of wafer polishing is to plane, flatten and polish the wafer in addition to removing metal deposits.

7.3.4 Electronic Chemicals – Market Size and Growth

The range of chemicals, technologies and supplier companies in the electronic chemicals segment is extremely broad. *Table 7.1* shows the estimated consumption of the major product categories in each of the principal geographic market areas in 2003, considering only chemicals used in fabricating semiconductors (front-end) and excluding those used for making printed circuit boards (PCBs) and other components (back-end).

Table 7.1. World consumption of electronic chemicals in 2003 (US$ million) according to *Davenport/Fink/Yoshio* (2004)

	United States	Western Europe	Japan	Rest of World	Total
Silicon wafers	1,400	885	2,700	1,015	6,000
Polycrystalline silicon	135	0	411	na	546
Photoresists	325	115	205	na	645
Atmospheric & specialty gases	585	305	514	na	1,404
Wet processing chemicals	370	170	346	na	886
Thin film metals (metal sputtering)	250	90	235	na	575
CMP slurries	230	80	105	na	415
Others (a)	435	82	248	na	765
Total	3,730	1,727	4,764	3,251	13,472

(a) III-V compound semiconductor wafers and low-k dielectrics for all regions

In 1998 about three quarters of the world market for electronic chemicals was concentrated in the United States, Western Europe and Japan. This dominance has rapidly eroded as the nations of Southeast Asia (including China) emerge as the leading market. Since that period, Taiwan, the Republic of Korea, Singapore, Malaysia and China have joined Japan and the United States as significant sources of technology and capital. These nations in 2003 held about 38 percent of the world market for overall electronic chemical consumption; they consumed over half of the materials used for the manufacture of PCBs (*Davenport/Fink/Yoshio* 2004).

Fig. 7.4 displays the predicted growth of various electronic chemicals market sub-segments in the near future. The highest growth rates in the electronic chemicals market are associated with advanced materials like low-k materials or CMP slurries, which have a huge impact on the manufacturing of advanced generation microchips.

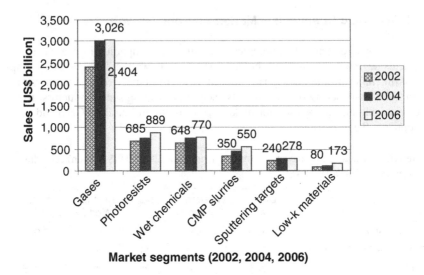

Fig. 7.4. Growth prediction for various electronic chemicals market sub-segments (2002 – 2006) according to *Chemical Week* (2003)

7.4 Semiconductor Industry to Get Boost in China

7.4.1 Demand of Integrated Circuits on the Rise in China

China is quickly outpacing some of the world's largest manufacturing centers.[1]

Consequently, China's consumer electronics industry is expected to post US$ 49.6 billion in sales revenue in 2004, growing by more than 20 percent annually to reach US$ 94 billion in 2007. Manufacturers are shifting to high-end products to differentiate themselves from the low-end competition.[2] Chinese manufacturers are employing more advanced technologies in various major product segments such as color TVs, DVD players, home theater systems, in-car entertainment, digital still cameras, MP3 players, personal digital assistants (PDAs) and household appliances.

---

[1] According to China's Ministry of Information Industry, 50 percent of cameras, 30 percent of TVs, 30 percent of air conditioners, 25 percent of washing machines, and 20 percent of refrigerators are produced or assembled in China. Databeans estimates that China also produces well over one third of global mobile phones, 17 percent of digital cameras, 20 percent of DVD players, and 28 percent of broadband routers and modems (*Electronic News* 2004a).

[2] This is reflected by China's integrated circuit consumption: In 2003, unit sales of ICs for digital consumer products grew by more than 22 percent, but ICs for traditional products, such as electronic toys and watches, rose by less than 14 percent (*Global Sources* 2004).

Nowadays, China plays not only a vital role as a manufacturer, but also as a consumer of electronics since the Chinese domestic demand for electronics is growing rapidly.[3] As a result of this development, the IC demand of the domestic Chinese electronic industry rose more than 40 percent in 2003. The market reached a value of US$ 25 billion (20.5 percent of the world's consumption) (*Business Daily Update* 2004a).

In 2003, China produced 12.4 billion units of integrated circuits (IC) and has become a global base for the production of semiconductor-based electronic products. However, domestically produced semiconductors only meet 20 percent of the country's demand for semiconductors (*Business Daily Update* 2004b).

Forecasts regarding the future development of Chinese IC demand vary significantly. Industry experts such as CCID Consulting, a Beijing-based IT consulting firm, expected the Chinese market to reach US$ 90 billion in 2008, accounting for 33.5 percent of global semiconductor revenue (*Business Daily Update* 2004a).[4]

7.4.2 China Emerging as Major Chip Manufacturer

This growing demand has attracted many global giants to set up plants on the mainland to gain a slice of the cake. Chinese chipmakers are also becoming increasingly competitive. Almost US$ 10 billion were invested in chip production joint ventures over the three-year period between 2001 and 2003, almost three times the amount invested in the previous three decades. A further US$ 5 billion has been pledged for investment (*RDSL Asia* 2004).

From 1999 to 2003, the production capacity of the China's IC industry expanded four-fold, averaging 45 percent annually, the fastest growth in the world in this period. Accordingly, China's share (as a producer) in the global IC industry rose from less than one percent to three percent.[5]

---

[3] According to the Ministry of Information Industry, cellular phone sales in China will reach 70 million sets in 2004, Chinese mobile users will reach more than 320 million. In the fourth quarter of 2003, the country's personal computer sales rose 22 percent to 3.9 million units from a year earlier, according to International Data Corp (*Business Daily Update* 2004a).

[4] This is in line with the prediction of the largest European chipmaker ST Microelectronics (STM), which forecasts that the market could be as large as US$ 100 billion in 2008 (*Dow Jones News Services* 2004). Other industry experts, like IC Insights or Databeans, however, are not so enthusiastic - but still optimistic - predicting a demand worth US$ 40 billion in 2007 (*South China Morning Post* 2004a) or US$ 69 billion in 2009 (*Electronic News* 2004a), respectively.

[5] The impressive growth story of the Chinese chip manufacturing industry in the very recent past can be illustrated by the following figures. According to data released by China's National Bureau of Statistics, China produced 9,418 million units of semiconductor integrated circuits (ICs) in the first half of 2004, up 43 percent year on year. From this local production, China exported 7,852 million units of ICs and microelectronic devices worth US$ 4.9 billion between January and June, up 51 percent and 83.6 percent year on year, respectively. According to China's General Administration of Customs, China's imports of integrated circuits and microelectronic devices in the period from January to June reached 29,272 million units valued at US$ 26.5 billion, up 45.3 percent and 65.0 percent year on year, respectively (*Asia Pulse Businesswire* 2004).

The potential for semiconductors over the next decade is similar to the growth of semiconductor manufacturing in Taiwan over the last decade, which went from a market worth US$ 10 million in 1990 to US$ 3.5 billion in 2004. And China as a market is ten times larger than Taiwan (*Venture Capital Journal* 2004). According to World Semiconductor Trade Statistics, China will cover 8 percent of global microchip production in 2008.

7.4.3 Drivers for the Chip Industry

Where New Fabs Are Built?

In a challenging and highly competitive global environment, successful chipmakers must endeavor to meet demand for their products, whenever and wherever that demand may arise. Therefore, semiconductor companies often position their chipmaking facilities, or fabs, in parts of the world where demand for their chips is high. Although many notable exceptions to this rule exist, the various regions of the world often play host to facilities that produce chips for the most heavily demanded end uses in those respective regions. Many factories that develop and produce cutting-edge chips for computers and high-end communications networks are located in the United States. Similarly, Asia/Pacific and Japan host a large number of fabs and foundries that focus on producing computer memory chips and other peripheral components that can be shipped to nearby contract electronics manufacturers (CEMs). Often, fabs located in Europe produce chips used in communications and consumer electronics products.

This rule also applies to the semiconductor companies with large international footprints. Intel, for example, focuses the majority of its microprocessor manufacturing operations in the United States, although the company also has less-extensive wafer CPU fab operations in Ireland and Israel. Intel's components assembly and testing facilities, on the other hand, are located predominantly in Malaysia, the Philippines, Costa Rica, and China.

Semiconductor companies that produce chips for widespread international distribution – as opposed to those that primarily serve regional demand - regularly locate their facilities in countries with inviting regulatory and tax structures, in addition to lower-cost labor pools. In situations where an overseas factory appears financially viable, U.S. and European manufacturers will often choose to build fabs in Asian nations that favor the semi-conductor industry, including Taiwan, Singapore, Korea, and recently China. Chinese government incentives such as free land and tax holidays have helped fuel the Chinese chip boom. China is increasingly becoming a location of choice for chip finishing operations (assembling, testing) because of lower operational costs there. Labor, electricity, power and other such expenses can be a substantial percentage of the overall outlay for testing and packaging facilities.

By contrast, operations that actually process silicon (wafer fabrication) involve expenses related to building the physical plant and buying equipment. As a result, local tax breaks are a larger factor than labor costs in determining where to build, and many manufacturers erect fabrication plants in relatively costly labor markets. AMD, for instance, is currently building its next fab in Dresden, Germany.

Foundry Boom

Another factor driving IC industry growth in China is the outsourcing of wafer production by semiconductor manufacturers. The 1980s saw the beginning of a trend toward outsourcing semiconductor manufacturing operations. Some firms chose to focus their energies on designing innovative chips, while relying on contract manufacturers, or foundry partners, to produce their products. Since they do not own or operate semiconductor fabrication facilities (fabs), these chip companies are referred to as "fabless".

But even chipmakers with fabs, often called integrated device manufacturers (IDMs), increasingly depend on foundries to ease some of the burdens of manufacturing. Many major semiconductor manufacturers are turning to foundries for chip production, including Motorola and STMicroelectronics, both of which have indicated a goal of outsourcing over 20 percent of their total chipmaking operations. IC Insights forecasted contract chipmaking sales to reach US$ 39.5 billion by 2008, up from US$ 5.5 billion in 1998 (*South China Morning Post* 2004b).

Asia is the locale for the major chip foundries. Taiwan leads the pack with foundry giants Taiwan Semiconductor Manufacturing Company (TSMC) (2003 sales: US$ 5.85 billion (*Reed Electronics* 2004)) and United Microelectronics (2003 sales: US$ 2.74 billion (*Reed Electronics* 2004)). Shanghai-based SMIC, founded in 2000 with investment from Singapore, the United States and the Shanghai government, is racing to become a world-class chip producer. The company manufactures memory and other basic kinds of chips. In 2004, SMIC (2003 sales: US$ 365 million (*Reed Electronics* 2004)) will storm past Singaporean rival Chartered Semiconductor (2003 sales: US$ 725 million (*Reed Electronics* 2004)) to become the world's third largest foundry, able to process the equivalent of 125,000 (200 millimeter) silicon wafers a month compared with Chartered Semiconductor's output of less than 110,000 (*Financial Times UK* 2004a).

7.4.4 New Fabs in China

China has 56 wafer fabrication plants in operation and 12 more under construction (*San Jose Mercury News* 2004a). Only eight of the 56 fabs operating in China have 8-inch wafer capability, while 17 fabs are still running 3-inch wafers, 15 use 4-inch wafers, six use 5-inch wafers and 10 use 6-inch wafers. SMIC is planning to build three 12-inch wafer fabs in Beijing. China had more than 400 chip design houses in 2002, and this number will grow to 500 next year. China also now has 108 IC chip assembly and test plants (*Electronic Engineering Times* 2004).

International semiconductor companies are currently following different strategies by erecting production capacities in China. Lucent Technologies, Motorola, NEC, Philips and Infineon have wafer-processing facilities in China. Infineon announced in 2003 plans to invest more than US$ 1.2 billion (around 30 percent of

its global investment) in China within the next three years (Frankfurter Allgemeine Zeitung 2004). European chipmaker ST Microelectronics is planning to establish an advanced fab in China by about 2005.

However, most companies such as industry leader Intel are investing in technically less demanding packaging and test plants.

The most advanced plants in China are currently erected by Taiwanese companies: In May 2004, Taiwan Semiconductor Manufacturing Company (TSMC), the
largest chip foundry in the world, was allowed by Taiwanese authorities to build
an 8-inch wafer plant in Shanghai. TSMC, in which the Taiwan government is a
major shareholder, was allowed to build this plant in China only after its 12-inch
wafer plant in Taiwan had been in production for six months. This policy is designed to keep the latest technology in Taiwan.

China is making huge strides in this key industry. As the manufacturing and design abilities of Chinese chipmakers improve, its semiconductor industry could
someday compete with companies like Intel. The first 300-millimeter wafer foundry in mainland China, run by SMIC began operation in September 2004, making
advanced DRAM chips for Infineon and Elpida using 0.11-micron and 0.1-micron
technologies. Industry experts had already speculated that SMIC would start the
trial production of wafers using the 90-nanometer technology and it will begin
mass production of SRAM chips using the technology in 2005. SMIC plans to
build three 300-mm plants in Beijing (*Taiwan Economic News* 2004, *China Post*
2004).

Even with the current aggressive expansion plans of SMIC and other wafer fabs
in China, local wafer production will still fall far short of domestic demand. As for
Shanghai-based foundry SMIC, by the end of 2004 it will have capacity for
125,000 wafer starts per month, representing only 5 to 6 percent of domestic supply. 40 percent of the company's wafer output was for overseas-based customers
(*Electronic News* 2004b).

7.5 China Promises Great Potential, but Plenty of Pitfalls

Signs of the tech boom are seen throughout China. Semiconductor factories are
rising on former farmland. High-tech districts brimming with start-ups, as well as
universities, are replicating the Silicon Valley model in China's sprawling cities.

Silicon Valley is contributing to this development. Intel, which employs 2,400
workers throughout China, will have invested more than US$ 1 billion there by the
end of 2005.

And there are signs of a Silicon Valley brain drain, as many Chinese-born entrepreneurs leave California to stake a claim in China's tech boom. Moreover,
China's abundant technical brainpower - about 220,000 bachelor's degrees in engineering were awarded in 2002, compared with 60,000 in the United States (*San
Jose Mercury News* 2004b) - is enabling the country to move into high-tech design
and development work. And the Chinese government is on a mission to make information technology a pillar of the national economy. It has targeted semiconduc-

tors and software, introducing incentives like cheap land and tax breaks for new companies.

Despite the rosy outlook, one has to note that there are risks in the capital-intensive semiconductor industry. A state-of-the-art plant costs up to US$ 2.5 billion and every new generation of fab doubles in costs. According to industry experts, Chinese chipmakers like SMIC would have to continuously invest in order to keep up with the technology to produce ever smaller and faster chips, resulting in heavy capital expenditures and leaving it dependant on external financing.

And despite the recent growth rates, the Chinese IC industry is still at the infant stage with great development potential. First, China is lagging in process technologies, IC design capabilities, and other critical areas. Another problem facing China is that it is restricted from having advanced chip technologies for ICs below 0.25-micron design rules under U.S. export laws that protect lithography processes and scanner tools. China is pressing the U.S. government to relax the lithography tool export laws, but it is unclear if or when that will happen. Consequently - up to now – U.S.-based chipmaker Intel does not have a silicon wafer production plant (the most sophisticated production work in the chip industry) in China, but only chip assembling and testing facilities.

In addition, foreign companies often feel that they can protect their technology much better if they keep manufacturing out of China. Concerns over theft of intellectual property and political risks are significant factors in dissuading microchip companies from investing in manufacturing or design in China. Of 18 U.S. and U.K.-based companies specializing in semiconductors and related devices, which gave their views in an informal poll, one third said worries about the potential loss of important technologies through intellectual property violations might inhibit them, or had already done so, from setting up new production or research operations in China (*Financial Times UK* 2004b).

Moreover, the intrinsic advantages of building a fab in China can be overstated. Labor costs are a small part of chipmakers' bills and highly-trained engineers have to be brought in. Utility costs are often higher in China than, for example, Taiwan. The lack of established equipment and materials suppliers adds further to costs. And tax incentives offered to advanced chipmakers are under threat. Chinese authorities' determination to attract chipmakers - offering not only free land and tax breaks but also direct investment and cheap loans from state banks - also carries risks. Many observers question the wisdom of making technology-intensive, labor-light semiconductors a policy priority in a developing country with endemic unemployment.

7.6 Electronics Chemicals in China

7.6.1 Market for Electronic Chemicals in China

Up to now, no detailed market reports addressing the various markets for electronic chemicals in China have been published. *Table 7.2* summarizes a consensus view of industry experts regarding the current and future demand of selected electronic chemicals in Mainland China, based on the assumptions that China accounts for 3% of the IC manufacturing output worldwide in 2003 and 8% in 2008:

Table 7.2. Consumption of selected electronic chemicals in Mainland China in 2003 (US$ million) and 2008

	2003	2008
Photoresists	8	50
Atmospheric & specialty gases	13	65
Wet processing chemicals	8	40

The future development of market segments such as low-k dielectric and CMP slurries, on the other hand, will be heavily dependent of the erection of advanced fabs in Mainland China which is still extremely difficult to estimate.

7.6.2 Strategies for Market Entry in China

In general, due to the industry dynamics, early entrants to the Chinese market are likely to benefit most from the anticipated growth. "Timing" is an essential issue. The appropriate strategy has to be selected, answering questions such as when to offer local technical service, start a local production plant or to open an R&D center in China.

General Considerations

Significant characteristics of the electronic chemical business are highly cyclical markets, continuing high R&D costs driven by rapid technological change, and a high level of technical competition leading to limited opportunities for product differentiation. In these respects, the electronic chemical industry resembles its customer, the electronics industry, more than the chemical industry.

The general strategy for designing manufacturing facilities for electronic chemicals should emphasize flexibility in production volume to minimize costs of idle capacity during downturns. Thus, numerous small manufacturing units, added incrementally as sales volume grows, are generally preferable to one large world-scale unit for any electronic chemical. Multipurpose equipment is common. The above-average obsolescence rate for electronic chemicals also dictates flexibility for such facilities. Consequently, establishing local production in China in an early phase is a reasonable strategic option.

The application technology for electronic chemicals is still largely in the domain of the customers rather than the suppliers. Technical progress has generally been driven by technical developments made by electronics companies rather than by innovations from the chemical companies. The emphasis in the R&D programs of materials suppliers is (and will remain) on technology acquisition and partnership with top-class electronic research organizations and R&D cooperation with top-innovative customers. Unless China has no "center of excellence" for the development of advanced microchips and the landscape of Chinese chipmakers is still dominated by foundries, which are not focused on their own development but instead rely on licensing in technologies, there is no need to establish R&D centers in China.

All players in the electronic chemicals market must place a premium on close and responsive relationships with the customer, literally and figuratively. These objectives are achieved by improving delivery times and by helping to maintain the purity of the product through solutions ranging from chemical recycling/replenishment to the operation of in-(fab)house distribution, just-in-time analytical service and regeneration systems and on-site chemical generation systems. Consequently, all manufacturers of electronic chemicals have to offer technical assistance to their customers in order to establish a significant business in China.

Beside these general considerations, individual aspects have to be considered since the overall market segment electronic chemicals can be subdivided into different sub-segments, each with its own characteristics. Manufacturers of photoresists will definitely select a different strategy than suppliers of wet chemicals in order to enter the Chinese market.

Strategies for Specialty Chemicals

Photoresists, CMP slurries and low-k dielectrics can be considered as pure specialty chemicals. Due to the high importance of product innovation, increasingly expensive R&D that addresses issues such as improving the properties of materials is also extremely important to the long-term success of chemical suppliers attempting to keep up with the performance demands of the semiconductor industry. The need for R&D has been and will continue to be a serious financial challenge to participants in these market segments of the electronic chemical business. Photoresists, CMP slurries and low-k dielectrics are priced on "value-in-use" and not on cost to manufacture, yielding relatively high operating margins.

Currently, all manufacturers of photoresists, CMP Slurries and low-k dielectrics still have their production and R&D facilities in those regions which are home to their traditional and established markets. And even for the next three to five years – unless China's share in the global production of microchips is less than 10 percent – there will be no need for these companies to establish local production plants in China. The costs of transportation contribute only to a tiny degree to total costs and the Chinese market can be served from existing plants.

Strategy for Commodity Chemicals

On the other hand, wet chemicals and gases can be characterized as semi commodities. Wet chemicals like sulfuric acid or phosphoric acid are generally considered as commodities. Nevertheless, ultra-high-purity (UHP) levels in wet chemicals (etchants, solvents) and gases are required for achieving adequate yields in high-density ICs. Particulate contamination has been found to account for 70 to 90 percent of all yield losses in IC manufacturing, while ionic contamination causes irreproducibility in semiconductor junction properties. These UHP purity levels require high-reliability process technology at the purification, packaging, distribution and point-of-use stages as well as advanced capabilities in analytics. The related industries have a number of significant barriers to entry such as the need for localized production facilities and secure raw materials supplies, long qualification processes with customers and detailed and long-standing quality expertise and know-how. These barriers keep this growing market secure for the incumbents and make it difficult for new entrants to enter. Due to the comparatively high costs of transportation, suppliers of these products are currently establishing local production facilities in China.

Part II: The Pharmaceutical Industry in China

8 China's Pharmaceutical Market: Business Environment and Market Dynamics

Michael Brueckner and Marc P. Philipp: Accenture GmbH, Kronberg, Germany;
Joachim E. A. Luithle: Bayer AG, Leverkusen, Germany

8.1 Introduction

Since the transition from a planned to a market economy started in 1978, China's economy has grown more than 8 percent per annum to a gross domestic product (GDP) of US\$ 1 trillion in 2003 – roughly the size of the United Kingdom's GDP (*Beckmann* 2004).

A closer look at China unveils a country faced with great complexities, driven by an enormous heterogeneity of customers, suppliers, competitors, regions, and government entities. Approximately 60 percent of the economy is represented by the country's east coast provinces; however, only 30 percent of the population lives there.

Not surprisingly, China's market attractiveness for multinationals varies greatly by sector. Whereas for some sectors China is the dominant force in driving global demand, other areas are still locked out from global trade due to regulatory issues or consumer preferences. With regard to the pharmaceutical industry, several multinational companies have already made significant commitments to shaping the Chinese industry – with some players starting as early as 20 years ago (*Prahalad/Lieberthal* 1998).

Despite limited short-term returns, major pharmaceutical players are pursuing aggressive growth strategies and trying to benefit from the Chinese market in the longer term. These efforts cover the entire value chain, including sourcing of active ingredients, research and development, and the production and selling of generic and proprietary drugs. In this, administrative hurdles, low healthcare spending, the lack of intellectual property protection and the poor distribution network infrastructure remain the biggest challenges (*Trinh* 2004).

This article outlines and analyzes the overall business environment, the healthcare system as well as market characteristics and dynamics for pharmaceutical multinational companies in China.

8.2 Business Environment for the Pharmaceutical Industry in China

8.2.1 Sociodemographic Business Environment

China's trend towards a more developed and affluent society has already initiated a significant shift in its demographic profile. The Chinese population is beginning to age – foremost in the urban areas (see *Fig. 8.1*).

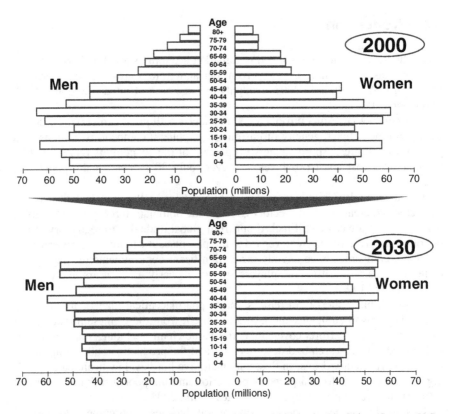

Fig. 8.1. Development of age distribution from 2000 to 2030 in the PR China (Source: U.S. Bureau of Census)

In urban China, the population aged 20 to 39, which accounts for most of the consumer spending, is expected to fall from 34 percent in 2002 to 30 percent in 2012. At the same time, it is expected that the proportion of the total population aged 65 and above will increase to 8 percent by 2008 and to 13 percent by 2025, a proportion that is even higher in major urban areas (*Liu/MacKellar* 2001).

These shifts, as well as the value which the Chinese place on good health, are likely to fuel demand for certain drug categories perceived to have preventive qualities, such as certain OTC pharmaceuticals, but also substitutes such as nutraceuticals or traditional Chinese medicine (TCM).

China's future disease profile is expected to remain more in line with developing countries than with the Western world, with cancer and heart disease accounting for relatively few cases compared with infectious diseases for the population as a whole. Statistics for the main diseases suffered by hospital inpatients reveal the divide between the urban and rural populations.

One of the major disease issues facing China in the future will be the increasing HIV epidemic. Some 840,000 official cases of HIV infection were reported in 2003. Nevertheless, one must assume underreporting especially in rural areas, understating the true extent of the problem. The projections of the National Intelligence Council (NIC) foresee as many as 10 to 15 million HIV/AIDS cases in 2010. However, these individuals will remain diffused among very large populations. The economic effect will be limited – at least until 2010. China has started to counter this problem with its 'China Plan of Action to Contain, Prevent and Control HIV/AIDS (2001-2005)' program.

Diabetes is another disease that might reach unknown dimensions in China. Diabetes being a multi-factorial disease, in which the genes not only interact with each other but also with environmental factors, genetically predisposed individuals will not necessarily develop Type 2 diabetes unless they are also exposed to environmental factors. Considering the effects of the demographic shift and of increasing affluence, the number of diabetes patients in China alone might exceed 150 million in 2025. The economic burden of the diabetes epidemic could be tremendous, if direct healthcare costs are considered along with indirect costs related to co-morbidity, mortality and loss of productivity. Hypertension is one of the most common co-morbidities linked to diabetes.

8.2.2 Economic Business Environment

Although China is one of the largest sources of overseas profit for many multinationals across all industries, they are beginning to feel competitive pressure from local market players. Chinese companies have significantly improved their capabilities and are increasingly able to pull together great market offerings. One reason is that domestic firms benefit from an enormous pool of workers, flexible working conditions, and an average hourly pay of approximately US$ 0.50 – even lower than Mexico's US$ 2.30 and India's US$ 0.80. Also, raw and processed materials are often cheaper in China and frequently match global quality standards; overall procurement costs generally run at 20 to 40 percent below the costs of comparable goods in established markets (*Joseph/Schaefer* 2004; *China Economic Information Network* 2003).

Despite great sourcing advantages, multinational drug makers' aversion to risk has resulted in imports being one of the driving forces, making up 40 percent of the Chinese pharmaceutical market in the mid 1990s (*Easton* 2003; *IMS Health* 2004). However, Chinese political leaders soon realized the importance of having their own champion league players in the pharmaceutical sector. Today, although import quotas and licenses are not applied to medical products, biotech-based equipment, instruments and consumables and imported medical products are subject to import tariffs of 4.2 percent plus an additional 17 percent value-added tax.

Moreover, by introducing non-tariff barriers, such as a central reimbursement list or financial promotion of the TCM industry, policy makers have been actively fighting against extensive spending on imported drugs. Consequently, as the market has developed, the market share of drug imports has plunged to as low as 10 percent in recent years.

In addition to China's expected disease progression, a large portion of the country's future market attractiveness depends on how consumer spending develops. Although 10 percent of the total population still lives below the poverty line and the "middle class" with an annual household income of more than US$ 20,000 makes up only 4 percent of the population, the growth trend will provide these 52 million people with the additional, necessary purchasing power to spend on healthcare, Western drugs and medical equipment.

8.2.3 Political and Regulatory Business Environment

It is expected that the political focus of the new Secretary General of the China Communist Party (CCP), Mr. Hu Jintao, will settle on two major issues in the coming years: advancing China's integration in the global trade community and improving China's social security system and infrastructure (*Hein* 2004).

With respect to the first issue, China's accession to the WTO in December 2001 has initiated the next round in opening up the country. However, the magnitude of change to be expected varies by industry sector, with the greatest impact in financial services and retailing. The pharmaceuticals market was already quite competitive before the country gained WTO membership.

Intellectual property (IP) rights – perhaps the most important issue for foreign R&D-based drug makers in China – are often expected to benefit significantly from WTO membership. Unfortunately, China's amendable IP climate is not the result of an inadequate legal framework, since the country's IP law conforms to most international IP standards such as the Trade-Related Aspects of Intellectual Property Rights (TRIPS) agreement granting 20-year patent terms and six years of data exclusivity from the date of marketing approval. Rather, the problem lies in poor enforcement, resulting in an estimated annual loss in sales of US$ 800 million due to trademark infractions and patent infringements. Of particular concern is the increase in counterfeiting, which costs multinationals approximately 10 to 25 percent of annual sales. Nevertheless, China's WTO commitment has influenced the drug distribution system. The retail pharmaceutical sector was opened at the end of 2004, a particularly important step not only for OTC product manufacturers.

Despite the general trend toward opening the country to world trade and generating economic prosperity, a strong government relations program remains an important factor in success in China. As in other emerging markets, the state utilizes its influence on market access and business rights to shape the involvement of foreign companies.

With respect to the second issue, the country's ill-preparedness for dealing with the SARS crisis and the particular problems faced by rural populations have moved disease prevention to a high position on the government's agenda. In 2003 and 2004, close to US$ 1.4 billion were spent on improving the healthcare system and develop new centers and hospitals for infectious diseases. The aim is to have a system in place within three years which is able to effectively respond to emergencies. In this, a major role will be played by the State Food and Drug Administration (SFDA), which was founded in 2003 to replace the State Drug Administration (SDA) and consolidate some responsibilities of the State Administration for Industry and Commerce and the Ministry of Health.

In addition, the government has established the 'National High-Tech R&D Program', giving biotechnology the highest priority among its seven categories with funding estimated at US$ 850 million from 2001 through 2005.

8.3 Market Characteristics and Dynamics of China's Pharmaceutical Industry

8.3.1 China's Healthcare Environment

Healthcare and Drug Expenditure

With a compound annual growth rate (CAGR) of more than 13 percent, China's total spending on healthcare has increased dramatically, faster than the country's general GDP levels.

Total healthcare expenditure accounted for 5.4 percent of GDP in 2002, up from 4.1 percent a decade earlier. However, OECD countries still spend significantly more – on average over 8 percent of their GDP – on healthcare. Furthermore, China's per capita spending is still 98 percent lower than the OECD average (see *Fig. 8.2*).

With respect to geographical differences, healthcare spending is very much concentrated on the more affluent provinces. Urban residents account for approximately 80 percent of total healthcare spending. The cities of Beijing, Shanghai, Jiangsu and Zeijang, home to more than ten percent of China's population, already account for 25 percent of total spending. In comparison to per capita annual spending on healthcare in the United States (US$ 300), China's expenditure is still 30 times lower (US$ 10) (*Liu/Rao/Hsiao* 2003).

The geographical concentration of healthcare spending is also found with respect to drug expenditure. Consumption generally varies widely, with Beijing and Shanghai accounting for three times higher drug expenditure per capita than Tianjin (*Forney* 2003).

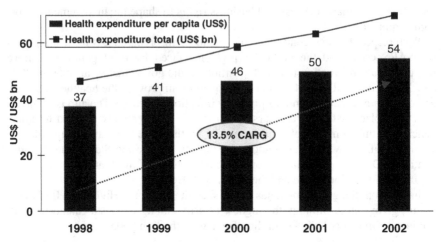

Fig. 8.2. Development of healthcare spending in the PR China (Source: Ministry of Health; Accenture Analysis)

Payers and Health Insurance System

Over the last decade, China's healthcare system has undergone various reforms as a result of cost explosions (since it provided basic healthcare services free of charge and thus established no incentives for cost effectiveness). In December 1998, the Chinese State Council agreed to establish a basic health insurance system for employees in the urban areas as part of the "Golden Social Insurance Scheme". For pragmatic reasons, the Ministry of Labor and Social Security (MOLSS) delegated the responsibilities for implementing the insurance scheme to regional level administrations. Consequently, only the major cities (again Shanghai is the leading force) have taken a professional approach to this initiative. Unfortunately, uncoordinated rudimentary systems still exist in many of China's provinces (*Lampton* 2003).

The "Shanghai model", currently China's most advanced health insurance system, consists of a welfare fund as well as individual and coordinated insurance accounts (see *Fig. 8.3*). However, as organizing the welfare funds is the responsibility of the city, provincial and regional administrations, a generalized view of the system is not practical.

Switching from a free healthcare system to one with a high level of individual financial contribution became unaffordable for the rural population, resulting in further aggravation. In 1998, 37 percent of the rural population could not afford a hospital stay; two thirds did not even have access to a nearby hospital. Currently, over 70 percent of the rural population pays for the major share of medical services entirely out-of-pocket.

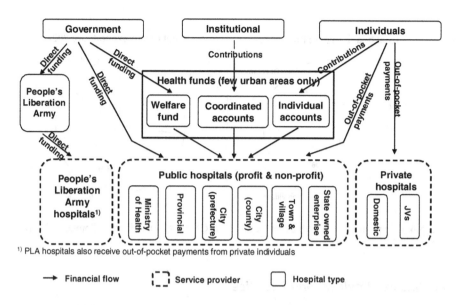

Fig. 8.3. Payers and providers in China's healthcare system (Source: Accenture Research & Analysis)

A program is currently underway to establish a healthcare funding scheme for the rural community. The ambitious goal is to cover all of China's 900 million farmers by 2010. The scheme will be funded by central and local governments, as well as by individuals, with subsidies promised for farmers living in the poorer central and western regions. The eastern province of Zheijiang is among the first to implement the new system and aims to have it fully operational by 2006.

Even the urban systems are still far from being perfect. In Shanghai more than 30 percent of the total population is still uninsured. As a consequence of the system transformation, individual contributions to healthcare expenses have nearly doubled (see *Fig. 8.4*). This trend toward private spending is likely to continue.

Nevertheless, by the end of 2003 more than 100 million citizens had joined the health insurance scheme. This has led to increasing demand for medical services; in 2002 China recorded over 2 billion hospital visits and almost 60 million inpatients.

The speed of improvement in the healthcare system is one of the key levers for providing a larger potential patient pool for reimbursed pharmaceutical products. However, increased consumer price consciousness due to out-of-pocket contributions has driven up the demand for generic products instead of expensive brands. Furthermore, the new health insurance system drives the dynamics of the hospital sector, which will have even wider implications on drug usage.

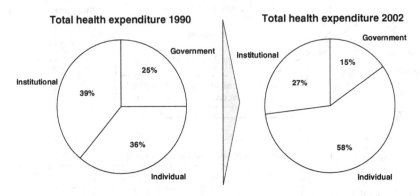

Fig. 8.4. Composition of health expenditure in PR China, 1990 and 2002 (Source: Ministry of Health)

Healthcare Providers and Facilities

The vast majority of medical benefits in China are provided through hospitals and other healthcare facilities. The total number declined by 7.4 percent in 2002, bringing it to as low as 310,000 facilities and reversing the growth trend. Contributing to the decline was a consolidation of health centers and outpatient clinics. Nevertheless, there was strong growth in the number of hospitals, which increased by more than 10 percent in 2002 to a total of almost 18,000 (see *Fig. 8.5*).

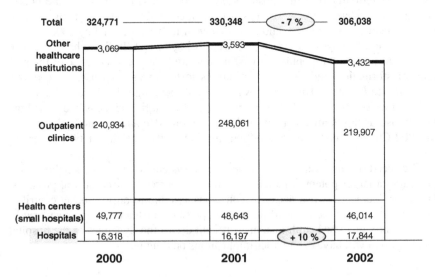

Fig. 8.5. Development in the number of healthcare facilities in the PR China (Source: Ministry of Health, IMS Health)

While almost half of healthcare facilities in China today are operated on a for-profit basis, the majority of hospitals (88 percent) remain non-profit institutions. There are currently very few private hospitals in the country, most of which charge fees comparable to U.S. facilities and thus primarily cater to expatriates. Selling drugs to these private providers requires a strong local presence, normally in the form of an agent or distributor.

The public hospital sector can be segmented into three tiers according to size and available equipment. Since the level of a patient's co-payments to the new insurance scheme differs according to the type of hospital, there is a trend towards polarization. Whereas patients with serious diseases prefer to go to tier III hospitals to receive the highest quality care, those with less serious conditions favor community hospitals (tier I).

Consequently, successful market penetration by pharmaceutical corporations will require further differentiation of sales channels for different product classes and target price levels. China's ongoing medical reform could result in stronger drug consumption. However, it will take several years for the reform to benefit foreign drug manufacturers.

Drug Distribution and Market Access Regulations

For the most part, drug manufacturers are not allowed to sell directly to hospitals and pharmacies but must go through state-owned wholesalers. Moreover, hospitals purchase most of their drugs through a tendering system.

Originally implemented to improve cost awareness, tendering has driven the general focus on price competition to such an extent that the drugs purchased through this system are often of very poor quality. The price discounts for hospitals achieved through tendering are often not passed through to the patient, but simply increase hospitals' profit margin. Furthermore, local manufacturers are still favored in tender processes since their investment in the local economy is regarded positively. Thus, bids are also evaluated according to the manufacturer's annual turnover and number of employees in the PR China.

Although it seems unlikely that the tendering system will be improved to deliver tangible benefits for patients in the short to medium term, the failure will support China's retail pharmacy sector (e.g. there is progress with the pilot scheme to separate hospital pharmacies from the hospitals).

In addition to the wholesale and tendering hurdle, hospitals carefully consider whether or not a patient can be reimbursed for his drug through the insurance scheme in their purchasing decision.

The patient is only refunded if a drug is listed on the insurance scheme's reimbursement list. This is based on the National Essential Drug List (managed by the SFDA), to which access is given only after a drug has been on the Chinese market for two years and has proven to be cost-effective.

To gain reimbursement status for their products, pharmaceutical companies should first focus their efforts on achieving provincial listing as this is regarded as an important factor in gaining national listing. Cost-containment measures will, however, include restrictions on the use of certain reimbursed drugs. Whereas two thirds of all reimbursed drugs are regulated with respect to their profit margin (mainly Rx products), one third are commonly used drugs (mostly generics) whose retail prices are regulated.

Governmental plans are now in place to establish a separate supplementary scheme to cover occupational medical expenses, which will include more expensive drugs than those on the standard reimbursement list. This may be a chance, especially for Rx drug makers, to make a deeper footprint in the Chinese market.

8.3.2 China's Pharmaceutical Industry Structure

The Chinese pharmaceutical market has demonstrated impressive growth since the mid-1980s, at a rate far exceeding annual GDP growth. Pharmaceutical sales in 2003 grew to US$ 6.2 billion, representing an increase of 20 percent over 2002 (see *Fig. 8.6*; *IMS Health* 2004). However, this expansive growth also reflects abnormal operating conditions driven by the SARS outbreak in 2003. Although China already holds a leading position for ethical pharmaceuticals in the region (exceeded only by Japan), the Chinese market is still comparatively small given the country's population. With an estimated average annual growth rate of 13 percent, the Chinese pharmaceutical market will achieve sales of over US$ 11.5 billion in 2008.

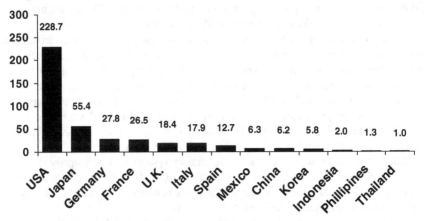

Fig. 8.6. Size and growth of key pharmaceutical markets in 2003 (US$ billion) – audited hospital market for ethical drugs (Source: IMS Global Market Report)

Market Structure and Segmentation

Chemical products account for around 70 percent of the pharmaceutical market in China. TCM (accounting for 24 percent) posts above-market growth and are particularly strong in rural areas (*Schmidt* 2002). Biopharmaceuticals make up 7 percent (see *Fig. 8.7*).

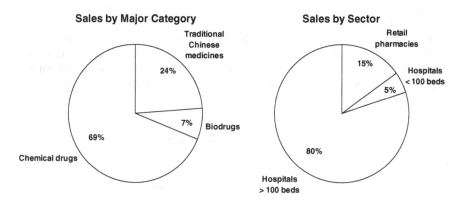

Fig. 8.7. Segmentation of Chinese pharmaceutical market by categories and sectors in 2003 (Source: IMS Health)

Approximately 85 percent of all products are sold through hospitals, with the majority of sales realized in the larger tier II and tier III hospitals (*Von der Hagen/Gruss/Wolff* 2002). The remainder is sold via retail pharmacies. The retail sector has grown significantly since the beginning of the new century and has currently more than 180,000 outlets and more than 800 pharmacy chains distributed unevenly across the country (e.g. Chongqing Tongjunge Pharmacy with more than 1,200 outlets) (*Schmitt* 2002a). Also, pharmaceutical manufacturers such as Shenzhen 999 or North China Pharmaceuticals have integrated vertically into wholesaling or retailing in order to increase channel power.

The Chinese market is dominated by generic drugs. In 2003 unbranded generics (products marketed under the generic name of their active ingredient) commanded 19 percent of the overall market. Original and licensed brands represented a combined 19 percent (see *Fig. 8.8*). Other products - mainly comprising branded generics, copycat products and products without licensing agreements - account for more than 60 percent of the market.

The local manufacturing sector specializes in the production of off-patent products such as antibiotics or copied drugs. Historically, new drugs in China were quickly copied and their price eroded by "illegal" generics. In the case of Pfizer's Viagra, four local counterfeits were already on the market before the product was actually introduced by its originator. As a consequence, multinationals have found it difficult to justify the investment to introduce innovative new products. It is estimated that another three years will be needed before intellectual property protection is enforceable, giving multinationals a bigger incentive to introduce innovative blockbuster drugs. In this context Pfizer, which is often quoted as an example of successful operations in China (having marketed seven out of ten blockbuster drugs in China), plans to introduce 15 innovative products. This is three times the number of products in the past five years and emphasizes the strategic and economic importance of this market. Furthermore, at the beginning of 2004 Pfizer established a strategic partnership with the SFDA bureau on anti-counterfeiting training, bringing global experience in this area to Chinese drug officials (*Sharper* 2002; *IMS Health* 2004).

It is estimated that around 98 percent of the drugs produced by domestic firms are replicas of foreign drugs. Government programs have been initiated (e.g. revised reimbursement lists) to reduce the price differential between off-patent original brands and generics, which may often be anywhere between 30 and 50 percent (*Vaishampayan/Chen* 2004).

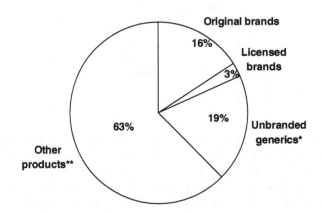

* incl. products marketed under the generics name of their active ingredient(s)

** incl. branded generics, copycat products, products where there is no evidence of a licensing agreement, products for which a licensing category has not yet been identified, as well as non-patentable products

Fig. 8.8. Segmentation of Chinese pharmaceutical market by licensing status, 2003 (Source: IMS Health)

Therapeutic Classes and Products in China

In 2003, local products generally posted stronger growth than those of multinationals, e.g. a JS Yangzijiang anti-infective grew more than 60 percent in the light of the SARS outbreak (see *Table 8.1*). Sales of this product were more than double those of the second largest product - Rocephin from Roche. With approximately 20 percent market share, systemic anti-infectives are the largest therapeutic area followed by cardiovascular, alimentary tract and metabolism (see *Fig. 8.9*).

Table 8.1. Sales of leading therapeutic products in China, 2003 (Source: IMS Health)

	Product (manufacturer)	US$ million [1]	Change (%)
1.	Zuo Ke (JS Yangzijiang Fty)	73	60
2.	Rocephin (Roche)	35	- 12
3.	Da Li Xin (Shenzhen)	34	79
4.	Heptodin (GSK)	31	- 4
5.	Sulbactam/ Cefoperazone (HLJ Harbin)	29	49
6.	Tienam (MSD)	27	- 1
7.	Glucobay (Bayer)	26	21
8.	Sandostatin (Novartis)	26	16
9.	Lu Nan Xin Kang (Shandong Lunan)	26	15
10.	Kai Shi (Beijing Taide)	23	38

[1] Ex-manufacturer prices MAT Q3 2003

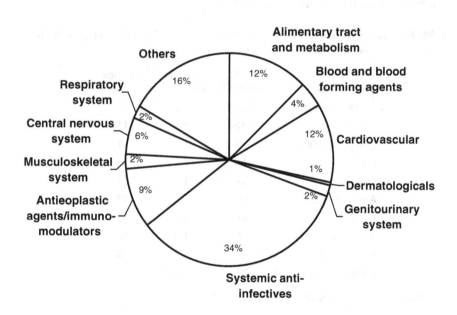

Fig. 8.9. Sales of leading therapeutic areas in China, 2003 (ex-manufacturer prices MAT Q3 2003; Source: IMS Health)

Competitive Environment and Pharmaceutical Manufacturers in China

The first multinational companies from Japan, the United States and Europe entered China to establish manufacturing, sales and marketing branches in the mid-1980s. They formed joint ventures with Chinese companies (Xian-Janssen) or established wholly-owned subsidiaries (Roche) to take advantage of China's low manufacturing costs and to sell into the Chinese market. Most multinational companies are located in the Golden Belt centered on Beijing and Shanghai.

Overall, companies that have established joint ventures or wholly-owned subsidiaries are regarded as performing more successfully than those that only engaged in import.

Nevertheless, domestic firms dominate the market. Today, domestic players hold approximately 65 percent of total sales compared to the foreign companies' share of 35 percent. In terms of volume, local drugs represent 85 percent of unit sales, while joint venture and imported products account for 9 percent and 6 percent, respectively. Recently, none of the leading multinational companies have achieved above-market sales growth. Jiangsu Yangzijiangs's anti-infective product helped the leading local company to dominate the hospital market, followed by Pfizer and GlaxoSmithKline (see *Table 8.2*).

Table 8.2. China's leading pharmaceutical corporations, 2003 – audited hospital market at ex-manufacturer prices (Source: IMS Health)

Corporation	US$ million[1]	Market share (%)	Change (%)
JS Yangzijiang Fty	147	2.5	46
Pfizer	113	2.0	11
GlaxoSmithKline	105	1.8	-1
Roche	105	1.8	5
Fresenius	89	1.5	12
AstraZeneca	86	1.5	12
Novartis	83	1.4	15
HLJ Harbin Pharm	74	1.3	26
Merck & Co	72	1.2	6
Johnson & Johnson	72	1.2	-1

[1] Ex-manufacturer prices MAT Q3 2003

In general, the majority of domestic companies are low-tech generic producers that mainly target the local market and compete almost exclusively on price or are bulk manufacturers exporting low-tech intermediates to unregulated markets. Since the late 1990s, some Chinese pharmaceutical companies have started exporting products to developed and unregulated markets to avoid strong pricing competition in their domestic market and to take advantage of marketing opportunities. As a consequence, Chinese companies now play a major role in the global bulk pharmaceutical and drug intermediary markets, thus accelerating pharmaceutical exports in recent years.

The Chinese pharmaceutical industry is seeing rapid consolidation as smaller domestic players exit the market due to their inability to comply with GMP requirements, which became mandatory starting in June 2004. According to information from the SFDA, the number of enterprises has dropped from 7,000 to 3,800. Companies that were not able to satisfy GMP requirements have been closed (*Feng* 2002).

The Chinese domestic industry is keen to work with Western companies in order to gain expertise in management techniques and to increase quality standards, with a view to competing on foreign markets. Among the most progressive is the Sanjui Enterprise Group, which acquired a majority stake in the Japanese company Toa Seiyaku. Even though the market perception is that multinational joint ventures and imported products will see their share of the market erode further, multinationals continue to emphasize their commitment to the Chinese market.

The true challenge for MNCs is not how to become profitable in China, but how to sustain a strong market position. China's growth and competition requires every market leader to continually reinvest its profits to maintain market share. Moreover, although profits can be earned and cash gained, adjusting the risk on those investments may likely change the picture.

Only a few domestic companies have the scale and capability to run effective R&D programs. Currently, domestic firms invest only a small percentage in R&D; commonly 2 percent of sales. However, the increased presence of multinational companies and especially the entry into the WTO have put pressure on the domestic industry to do more than counterfeit proprietary products. As a result, R&D is increasingly being encouraged by the Chinese government at both national and provincial levels. Domestic companies and research institutes increasingly invest in state-of-the-art technologies such as molecular diagnostics, stem cell research and pharmacogenomics. For instance, the Shanghai Institute of Materia Medica and the Beijing Genomics Institute are investigating the SARS virus. The Chinese government has clearly stated that it intends to become a major player in the international biotech industry.

Up until now, a limited number of multinationals who are represented in China by the R&D Based Pharmaceutical Association (RDPAC) have set up R&D centers (e.g. AstraZeneca, Novo Nordisk, Servier and Pfizer). Other multinationals are expected to follow including Roche, GSK and Lilly, which have set up research laboratories in Zhanjiang Hi-Tech Park – Shanghai's medicine valley (*Chuan* 2004; *Torreblanca* 2004).

Market for Over-the-Counter Drugs

The Chinese OTC market is evolving. In 2003, nearly 700 products were classified as OTC by the SFDA under the Pharmaceutical Administration Law, which for the first time actually established an official OTC category. Historically, OTC medicines have not been separated from prescription drugs and many still qualify for reimbursement. However, the government is phasing out reimbursement of these drugs. The official OTC product list includes chemical products as well as TCMs and consists of categories A and B. Category A comprises products that can only be sold in retail drugstores or pharmacies; category B are products that can be widely sold in general retail outlets and drugstores without pharmacy supervision. China's over-the-counter market is currently valued at over US$ 1.8 billion and is estimated to reach US$ 6 billion by 2010. Generally, quantification of the OTC market, which is very competitive and dominated by local and joint venture products, is difficult due to its complex distribution system and inaccurate reporting of retail sales (*Espicom Business Intelligence* 2004).

Traditionally, OTC products are sold through hospital pharmacies, with several OTC products included on reimbursement lists. The OTC market is growing well above the overall pharmaceutical market. This is primarily due to an emergence of retail pharmacies, the SARS outbreak, the growing health consciousness of the middle-class and limited access of the rural population to expensive Western medicine. A number of multinational companies such as GlaxoSmithKline, Johnson & Johnson and Novartis have positioned themselves to take advantage of this rapid growth. Bristol-Myers-Squibb's OTC business in China is large but declining against the trend.

Distribution System for Pharmaceutical Drugs

The distribution system for pharmaceutical products in China is highly complex and characterized by regional and local diversity. Drug makers are dependent on a loose network of domestic distributors and sub-distributors who operate on a city or regional level separated by extensive formal and informal trade barriers. Among the biggest obstacles are China's transportation system and storage facilities. Although the government has begun to improve this sector, widespread progress in this area will take years to accomplish. Reform of the distribution sector is gradually advancing towards the government's objective to create at least 40 national and regional distributors which will cover 70 percent of the market and replace mostly state-owned wholesalers operating at the local level (*Schmitt* 2002a).

China is experiencing an accelerated consolidation in the wholesaler sector with at least 4,400 wholesalers exiting in 2002 and more companies going out of business due to non-compliance with GSP. In this context the three leading distributors - CNPG, Shanghai Pharmaceutical Group and Guangzhou Pharmaceutical Group - are building regional networks and selling directly to hospitals and the retail sector.

The first foreign entrant into the distribution sector was Zuellig, which established a joint venture with the China Xinxing Medicine Company in 2004. The joint venture will achieve access to direct distribution (i.e. circumvent the need to sell via wholesalers as in the past). With the full opening of the retail sector to foreign investors at the end of 2004, more companies are expected to enter the market and create national networks, competing head on with the three leading distributors. Foreign distribution companies do not have geographical limitations but are only allowed to operate in conjunction with a domestic joint venture partner (*Vaishampayan and Chen* 2004).

In the retail sector three distinctive outlets are emerging: chain drugstores, discount pharmacies (estimated at 1,000 and selling mainly low-price generics) and pharmacies designated for reimbursement.

The changing dynamics in the distribution channels are driving pharmaceutical companies to adapt their sales and marketing strategies. Most companies have already established teams to professionalize their tender bidding for hospitals. While the larger tier II and tier III hospitals remain the focus for marketing new premium-priced drugs, Tier I hospitals have to be targeted as well in order to manage the increasing number of patients receiving prescriptions in these community hospitals. Moreover, the growing importance of the retail sector is also encouraging companies to establish a dedicated retail pharmacy sales force, concentrating on the Beijing-Tianjin, Shanghai Delta and Guangdong-Pearl River Delta regions.

8.4 Conclusion and Outlook

The Chinese pharmaceutical market undoubtedly represents one of the most important current and future markets for foreign small and medium-sized companies as well as for multinational enterprises. Many companies such as Bayer, Schering, GSK and Pfizer have already successfully established local operations in this strategic market and are currently expanding their commitment. The opportunities for foreign companies are promising and range from marketing and selling generic and proprietary drugs locally, sourcing of chemical substances, setting up local production facilities and conducting research and development.

However, these positive aspects cannot conceal the fact that, at the beginning of the new millennium, China is still a developing country facing typical challenges such as a high level of bureaucracy and protectionism, a lack of sophisticated infrastructure and underdeveloped healthcare and patent systems. These limitations often make it difficult for foreign companies to justify the huge investments necessary to build market share, local capabilities and footprint in this very competitive market.

9 Developing the Pharmaceutical Business in China – The Case of Novartis

Angela Wang: Beijing Novartis Pharma Ltd., P.R. China;
Maximilian von Zedtwitz: Tsinghua University, Beijing, P.R. China

9.1 Changing Pharmaceutical Environment in China

China's transition from an emerging to an advanced economy is eminent in the development of its pharmaceutical market. On the basis of a projected average annual GDP growth rate of 7 to 8 percent, China's share of the global pharmaceutical market will increase from about US$ 6 billion in 2002 to an estimated US$ 24 billion by the end of 2010 (see *Table 9.1*, BCG analysis). China has the world's fastest-growing over-the-counter (OTC) drug market and is now the second largest pharmaceutical chemical producer.

Various reform policies are affecting economic development in China, which has been under the new leadership of President Hu Jintao and Premier Wen Jiabao since 2003. Some of the most promising developments are China's entry into the WTO (2001), growing domestic consumption and investment, and the efforts made in preparation for the 2008 Olympics in Beijing and the 2010 World Expo in Shanghai. However, China still faces a strong imbalance in development between its eastern and western regions, a huge income gap between urban and rural areas, a mounting unemployment pressure and a sub-standard financial system, just to name a few.

Table 9.1. China is expected to become the fifth largest pharma market by 2010 (figures in US$ billion)

2002 Top 10		2005 Top 10		2010 Top 10	
USA	196	USA	262	USA	466
Japan	53	Japan	65	Japan	81
Germany	20	Germany	24	Germany	37
France	19	France	21	France	28
U.K.	14	U.K.	16	*China*	*24*
Italy	13	Italy	15	U.K.	24
Spain	9	*China*	*14*	Italy	23
Canada	8	Brazil	10	Canada	17
Mexico	8	Canada	10	Spain	16
China	*6*	Spain	10	Brazil	15
Total	**346**	**Total**	**447**	**Total**	**731**

The Chinese pharmaceutical market has been growing at annual rates between 8.4 and 15.5 percent since 2000 and is expected to grow at an average 11 to 12 percent annually in the coming years. This growth is driven in part by an increase in income levels and earning power, an overall improvement of the quality of life in (mostly eastern coastal) China, and the associated consumers' awareness of diseases and a desire for improved healthcare. As living standards rise, particularly in the cities, a number of formerly common diseases and conditions associated with poverty have been eliminated almost entirely. At the same time, higher incomes, new diet patterns, decreased physical activity and more work-related stress have combined to increase the incidence of diseases new to China such as diabetes, cardiovascular disease and other stress-related disorders. Healthcare coverage is becoming more widespread and a growing ageing population contributes to the demand for more pharmaceutical products. At the same time, the pharmaceutical industry is still hampered by a lengthy new product registration process, limited drug reimbursement listings and limitations in new drug administration law. These developments and the increasing affordability of products to the general population have yielded various opportunities for foreign companies, particularly in the retail market and in the development of new therapeutic areas.

9.2 Novartis Pharma in China

Novartis is a major pharmaceutical and healthcare company operating through 360 independent affiliates in 140 countries. In 2003, Novartis Group's sales amounted to US$ 24.9 billion with a growth of 19 percent. Its profits added up to US$ 5.9 billion with growth of 16 percent and its profit rate reached 23.7 percent. Pharmaceuticals sales climbed 18 percent to US$ 16.0 billion.

In China, Novartis is represented by Beijing Novartis Pharma Ltd. China (Novartis Pharma for short), a joint venture between Novartis Group headquartered in Basel, Switzerland, and two Chinese partners: the Beijing Pharma Group and Beijing Zizhu Pharma. This joint venture was established as the first pharmaceutical venture in Beijing in 1987 with an initial investment of US$ 21 million (a later expansion is planned to double this investment). By 2004, Novartis Group held a 78-percent stake in this venture. Novartis Pharma has about 1,400 employees in China, nearly all of whom are Chinese locals. Novartis operates one manufacturing center near Beijing with an annual capacity of 1 billion tablets and capsules and 12 million gel forms.

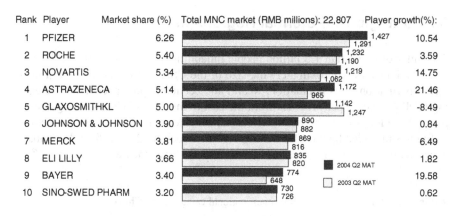

Rank	Player	Market share (%)	Total MNC market (RMB millions): 22,807	Player growth(%):
1	PFIZER	6.26	1,427 / 1,291	10.54
2	ROCHE	5.40	1,232 / 1,190	3.59
3	NOVARTIS	5.34	1,219 / 1,062	14.75
4	ASTRAZENECA	5.14	1,172 / 965	21.46
5	GLAXOSMITHKL	5.00	1,142 / 1,247	-8.49
6	JOHNSON & JOHNSON	3.90	890 / 882	0.84
7	MERCK	3.81	869 / 816	6.49
8	ELI LILLY	3.66	835 / 820	1.82
9	BAYER	3.40	774 / 648	19.58
10	SINO-SWED PHARM	3.20	730 / 726	0.62

2004 Q2 MAT
2003 Q2 MAT

Fig. 9.1. The top10 pharmaceutical MNCs in China (Source: IMS China Q2 2004)

Worldwide, Novartis is the fifth largest pharmaceutical company in terms of sales. Among the multinational pharmaceutical companies in China,[1] Novartis occupies the third spot with 5.34 percent in market share (in Q2 2004 MAT, see *Fig. 9.1*). Pfizer is currently market leader with a share of 6.26 percent, although AstraZeneca is the fastest-growing company at a rate of approximately 21 percent in 2003/2004. Total MNC market growth was 10.98 percent in 2003.

Novartis Pharma has maintained a compound annual growth rate (CAGR) of about 20 percent between 1999 and 2003, and is now the leading company in hypertension, oncology and transplantation. Among Novartis Pharma's recent product launches in China are Femara and Glivec in 2002, Zelmac, Trileptal, Starlix and Simulect in 2003, and Sandostatin LAR in 2004. Co-Diovan, Visudyne, Comtan and Zometa are expected to follow in 2005. The pipeline is rich and more product launches are scheduled. Novartis also uses one of the largest field forces in the industry with a consistent record of sales productivity. It is concentrated primarily in the strategic regions but is now also expanding fast in second- and third-tier cities in inland China.

Novartis Pharma is engaged in a number of initiatives designed to improve its corporate citizenship. It supports and sponsors various programs with ministries and official departments of the Chinese government, as for instance the Public Education Program on Organ Donation with the Ministry of Health, and a long-term educational program with the State FDA which it helped established. Also, it has been co-organizing National Medical Insurance symposia with the Ministry of Labor and Social Security each year since 2001. Novartis Pharma also produces Coartem, a drug to treat malaria, and supplies it to WHO at cost.

---

[1] Among the top 50 pharmaceutical companies in China, 32 percent are foreign companies or joint ventures. Foreign-funded enterprises accounted for 35 percent of market share while imported drugs account for 15 percent of the pharmaceutical market.

9.3 Strategic vs. Potential Regions in China

A country of more than 1.3 billion people inhabiting an area of nearly 9.5 billion square kilometers (approximately the size of the USA), China is far from a uniform distribution of its population, wealth and economic power. Given a decreasing GDP per capita per province from the east to the west of China, foreign companies have focused on the wealthier coastal provinces to develop their business.

Figures provided by the China Statistical Yearbook for 2003 reflect this imbalance between the east and west, also with respect to healthcare expenses (see *Table 9.2*).

Table 9.2. Vital statistics of all four main regional categories in China

Regions	Popu-lation (m)	GDP per capita (US$)	Income per capita (US$)	Healthcare expense per capita (US$)	Number of hospitals	Number of doctors per 1000	Number of visits (m)
Megapolis	119	2,270	1,074	54	1,802	1.85	267
Eastern	296	1,618	749	37	3,911	1.54	293
Central	659	824	447	22	8,593	1.33	422
Western	201	646	372	21	3,538	1.46	126
China	1,275	1,115	564	28	17,844	1.45	1,108

Given these different market characteristics, market development strategies need to be adjusted to fit the respective demands and opportunities in targeted Chinese provinces. Further analysis reveals that GDP growth per capita is nearly twice as high in selected eastern provinces compared with the western provinces, as is the average drug expenditure per capita. These strategically important eastern provinces include the major cities of Beijing, Shanghai, Tianjin and Guangdong, and the comparatively wealthy provinces of Shandong, Zhejiang, Jiangsu, Guangzhou and Fujian (see also *Fig. 9.2*). All of these are situated on China's east coast.

Even within certain provinces the average drug expenditure is not uniform, as is shown in *Fig. 9.3*. Beijing, Guangzhou and Shanghai still lead in terms of drug expenditure per capita among major Chinese cities, but other cities such as Hangzhou (Zhejiang province), Nanjing (Jiangsu), Jinan (Shandong), Wuhan (Hubei), Xian (Shaanxi) and Shenyang (Liaoning) are catching up quickly. Although Chongqing (Sichuan) still has a relatively low average drug expenditure ratio, it is of strategic interest for pharmaceutical companies as it is China's most populous city (30 million people) and hence relatively convenient to develop from a distribution point of view.

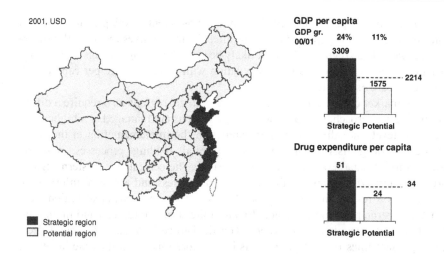

Fig. 9.2. China's eastern provinces are better suited for the development of the pharmaceutical market

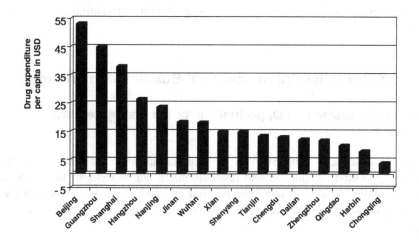

Fig. 9.3. Difference in drug expenditure per capita in major Chinese cities

The entry strategies of foreign pharmaceutical companies are thus focused on the richer regions in the east, with the intention to expand into central and western provinces in the future. This is motivated by the fact that the populations in these strategic regions can afford the more expensive ethical, modern and effective drugs, and the fact that they have higher incomes, better education and higher disease awareness. There is also a more advanced healthcare infrastructure in eastern China and physicians are more open to scientifically driven marketing. The strategic eight provinces and cities account for about 75 percent of the current prescription business in China. Subsequent development of central and western China is expected to follow after the initial investments in the strategic regions. These regions will most likely be served with older drugs that are off-patent and hence affordable for a population with lower average per capita GDP and healthcare expenditures.

Since market characteristics differ, these strategic regions also require a different business approach. Overall, the strategic regions are managed at a higher professional level, including more transparency in the implementation of the reimbursement drug lists (RDL) and the hospital bidding processes, and the availability of data about patients and diseases. The central and western regions are characterized by more relationship-driven business and access to information is poor. As a result, business support differs in those two principal regions: the marketing strategy is more science-driven in the strategic regions and more relationship- and price-driven elsewhere. The field force is organized around dedicated product lines in strategic regions but is more mixed and flexible in central and western China. Most of the field reporting is still collected manually, as opposed to electronic reporting in the developed strategic regions. Given different market and business characteristics, pharmaceutical companies thus focus on the eastern regions with more expensive innovative drugs, leaving the more mature and less expensive drugs for the developing markets in central and western China.

9.4 Analyzing the Pharmaceutical Business Environment

9.4.1 An Assessment of Opportunities for the Pharmaceutical Business

The typical business value chain in the pharmaceutical industry is defined by R&D, registration, production and distribution. In our analysis of the Chinese business chain, we also included the hospital and outlet intermediaries, and the patient reimbursement services as they are of critical importance in China (see *Fig. 9.4*).

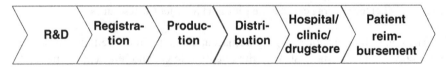

Fig. 9.4. The pharmaceutical industry business chain in China

China does not have a particularly strong reputation in researching and developing new molecular compounds and drugs for pharmaceutical use, although the Chinese government has declared this a major priority to close the gap with other developed economies. As of 2004, most of the class I (new drugs and new compounds) and class II (drugs already approved in another country) products were still developed abroad and sold by foreign multinationals in China. However, Chinese pharmaceutical firms are quite strong in the generics market, where they leverage their local roots and strengths in copying and marketing older drugs.

Local R&D presence usually begins with R&D collaborations. Glaxo-Smithkline (GSK), for instance, has invested over US$ 10 million in cooperative R&D with Chinese research institutions since the mid-1990s. Novartis has a collaboration with the Shanghai Institute of Materia Medica (SIMM), with the objective of isolating compounds from Chinese medicinal plants for Novartis to further screen and identify lead compounds. After an initial phase and US$ 2 million in funding, training and equipment, SIMM had by 2004 isolated more than 1,800 compounds from natural herbs covering immunology, oncology, diabetes and the central nervous system. With additional funding, SIMM is expected to isolate a further 1,500 compounds for Novartis over the next three years.

More recently, foreign pharmaceutical companies have also invested in the establishment of R&D centers in China (see *Table 9.3*). Compared to R&D efforts made in China by other industries, however, dedicated pharmaceutical R&D centers have been lagging behind, mostly because of concerns over protection of intellectual property and the low quality of R&D work. Also, initial pharmaceutical investments were focused on developing sales and production rather than R&D. Since R&D quality has been improving rapidly in China, several R&D centers have been established or are under consideration. Most of these R&D investments are development-oriented clinical R&D centers. Also, the U.S. FDA has accepted the use of Chinese clinical trial data for the submission of new drug approvals. Due to the huge Chinese population there are large numbers of patients available for all major genotypes, which makes clinical testing faster and more efficient. Together with the lower costs of R&D in China and the increasing need to enhance the public image of pharmaceutical companies in China, we expect the number of pharmaceutical and biotech-related R&D centers to increase in the near future.

The registration phase for a new product still lasts between three and four years in China, which is quite lengthy. Even for drugs approved previously elsewhere, the formal approval by Chinese authorities is far from certain, easy or transparent. Reform of the registration process has thus focused on reducing the time required. The previous five categories of new drugs recognized by China's drug approval system have been replaced by a classification based on the type of new drug application. The definition of new drug has also been changed to cover products which have never been marketed in China as opposed to the former definition of products which have never been manufactured in China. Also, the Chinese State FDA has promised to keep data submitted for drug registrations confidential until 2007.

Table 9.3. Foreign pharmaceutical R&D centers in China (Source: SMIE Medicine Information and own research)

Year	MNC	Name	Location	Investment	Objective
2001	Servier	Servier (Beijing) R&D Center	Beijing	n/a	Develop potential value for TCM
2002	Novo Nordisk	Novo Nordisk (China) R&D Center	Beijing	n/a	General biotech research
2003	Astra-Zeneca	East-Asia Clinical Research Center	Shanghai	First year: US$ 4 million	Clinical research
					Collaboration with Health & Medical Institute in China
					Localize therapeutic methods
2003	Eli Lilly	Shanghai Chem Explorer Co Ltd.	Shanghai	All funds supplied by Eli Lilly, > 100 scientists	Combining different kinds of organic substance for new drugs
2004	Roche	China R&D Center	Shanghai	First year several million, 50 scientists	Phase I: Chemical drugs; analyzing compound structures
					Phase II: TCM & genetic engineering
2004	GSK	n/a	Tianjin	16 scientists	Develop 20 new OTC products in next 3 years
2004	J&J	n/a	Shanghai	n/a	Develop medications suitable for Chinese and Asians

Changes are also under way with respect to production and manufacturing standards in China. It is expected that compliance with international GMP (good manufacturing process) will weed out about half of the more than 6,000 drug companies operating in China. About 1,700 of these are foreign investments and less than 1,000 are GMP-certified. More than 1,350 of the manufacturing drug companies are producers of raw materials. Particularly the Chinese manufacturers are very strong in their ability to copy foreign drugs, sometimes selling them under the foreign label. Almost 99 percent of the 3,000 pharmaceutical products manufactured in China are copies of foreign products, either legal generics or illegal counterfeits. But besides the ethical, mostly foreign-dominated drug market, more than 8,000 traditional Chinese medicines (TCM) are manufactured and sold.

China's drug distribution system is still fairly ineffective, which provides great opportunities for foreign companies which can establish national shipping and drug logistics networks. To date, China has no large-scale nationwide distribution system and relies on 17,500 often regional wholesalers, creating many layers between manufacturers and patients. Clearly, consolidation is needed.

80 percent of drugs are dispensed in hospitals, only about 15 percent in pharmacies (the rest is distributed through clinics). Drugs sold through pharmacies are typically less expensive than in hospitals. The growth of drug sales in more than 16,000 hospitals at county level or above amounts to about 11 percent, while growth in the approximately 120,000 pharmacies was almost 20 percent, mostly because more and more costs are being paid by the patient; a result of China introducing an individual insurance scheme that results in 'patient co-pay', increasing the cost awareness of the end consumer. A second reason, which further promotes the growth of pharmacies, is the progress of a nationwide healthcare reform, part of which is transforming hospital pharmacies into independent pharmacies, i.e., the separation of the service from selling medicine, and making hospital services a viable business in itself. Most Chinese hospitals are currently subsidized by their pharmacy businesses.

The patient reimbursement scheme is also tightly coupled with the reform of medical insurance. The new urban employee health insurance scheme, which replaces the free health insurance services for government employees and the labor health insurance system for employees of state-owned enterprises, covered 100 million people in 2003; the target for 2004 was 130 million. This new health insurance has started in the cities, as it is tied to employment and implemented as a deduction from a worker's salary. The neediest people are actually the farmers in less developed regions of China, as they have hardly any money to pay for medical expenses. However, they are not in an employee relationship and hence have no system or means of participating in this program. Pharmaceutical firms try to have their drugs included in the reimbursement drug lists (RDL) established at provincial level (generally – with some flexibility – following the recommendation of the national essential drug list or EDL). This process alone can take up to two years after product launch. Only those drugs included will be reimbursed partly or fully by China's insurance system. Other drugs will have to be paid for by the patients themselves.

9.4.2 Challenges for Pharmaceutical Companies

In summary, some of the major challenges that pharmaceutical companies in China are currently facing are:

- Healthcare reform
- Reimbursement limitation
- Price control
- Hospital bidding
- Rational drug use
- The new law on drug administration

All these measures are aimed at improving the healthcare system while limiting the costs of drug treatment. For instance, while it is clearly necessary to have drugs included in reimbursement drug lists, those drugs will then be subjected to price controls by the provincial or state governments. As an additional price control mechanism, drugs are also subject to a hospital bidding process a part of the hospitals procurement process. Although a variety of factors are considered in this bidding process, price is a very important one. It is not surprising that the multinational pharmaceutical industry is trying to convince the government to provide special consideration to original off-patent products by placing them into a separate innovative category, as important off-patent drugs have a different price structure from local generic drugs. As in other more developed countries, cost containment often goes hand in hand with a loss of healthcare quality.

With regard to rational drug use in China, in the past doctors often had the tendency to prescribe more drugs than perhaps absolutely necessary, partly in order to ensure effective treatment of a disease and partly perhaps due to heavy promotion by pharmaceutical companies. While this habit is deemed questionable in terms of actual long-term therapeutic efficiency, it also increases the costs for patients and healthcare. However, overprescribing benefits the revenues of pharmaceutical companies. Doctors are now encouraged to prescribe only what is absolutely necessary within their judgment, which is considered a financially and hopefully medically positive development for the healthcare system. Patients also need to be better educated about the risks of abusing or misusing certain drugs, as they lead to unwanted side-effects if not taken as prescribed. This effect is most dangerous in the case of antibiotics, where premature termination of the therapy results in resistant strains of bacteria that cannot be treated with existing antibiotics. Non-prescribed sales of antibiotics have now been made illegal in China.

The introduction of the new Pharmaceutical Administration Law in December 2002 was designed with WTO membership requirements in mind. Although this new law has brought some improvements along the pharmaceutical business chain in China, there are persisting issues of concern to multinationals relating to inadequate protection of intellectual property and restricted market access as a result of regulatory procedures and measures deemed to favor local manufacturers. These concerns have ultimately negative consequences for the smooth introduction of new drugs to China.

In terms of market share, the Chinese ethical market is dominated by systemic anti-infectives and drugs for alimentary tract, metabolism and cardiovascular indications. They constitute over 50 percent of the entire pharmaceutical market in China (see *Fig. 9.5*).

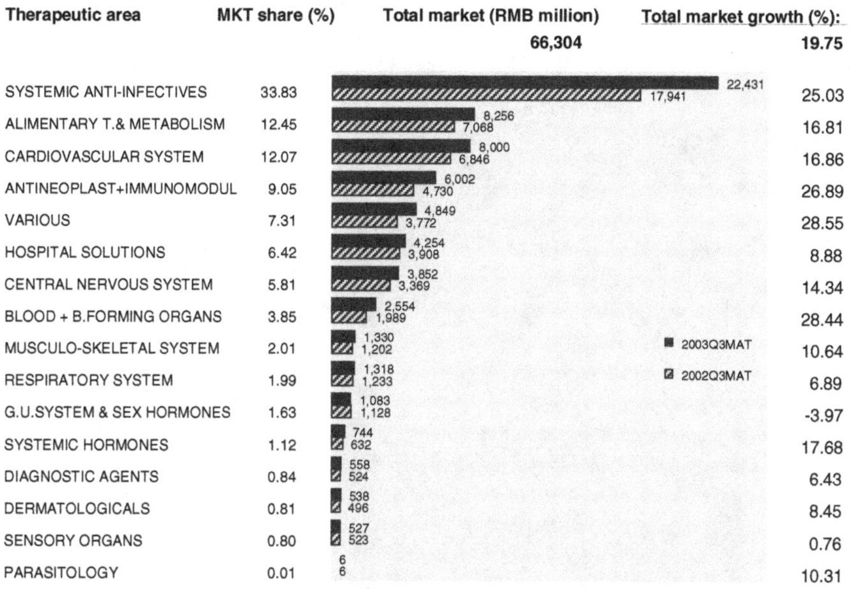

Therapeutic area	MKT share (%)	Total market (RMB million) 66,304	Total market growth (%): 19.75
SYSTEMIC ANTI-INFECTIVES	33.83	17,941 / 22,431	25.03
ALIMENTARY T.& METABOLISM	12.45	8,256 / 7,068	16.81
CARDIOVASCULAR SYSTEM	12.07	8,000 / 6,846	16.86
ANTINEOPLAST+IMMUNOMODUL	9.05	6,002 / 4,730	26.89
VARIOUS	7.31	4,849 / 3,772	28.55
HOSPITAL SOLUTIONS	6.42	4,254 / 3,908	8.88
CENTRAL NERVOUS SYSTEM	5.81	3,852 / 3,369	14.34
BLOOD + B.FORMING ORGANS	3.85	2,554 / 1,989	28.44
MUSCULO-SKELETAL SYSTEM	2.01	1,330 / 1,202	10.64
RESPIRATORY SYSTEM	1.99	1,318 / 1,233	6.89
G.U.SYSTEM & SEX HORMONES	1.63	1,083 / 1,128	-3.97
SYSTEMIC HORMONES	1.12	744 / 632	17.68
DIAGNOSTIC AGENTS	0.84	558 / 524	6.43
DERMATOLOGICALS	0.81	538 / 496	8.45
SENSORY ORGANS	0.80	527 / 523	0.76
PARASITOLOGY	0.01	6 / 6	10.31

Legend: ■ 2003Q3MAT   ▨ 2002Q3MAT

Fig. 9.5. Leading therapeutic classes in China (Source: IMS plus Q3 2003 MAT)

However, if measured by annual growth, other therapeutic areas come out on top (after the 'various' category): drugs for blood formation and blood-forming organs, antineoplasts and immunomodulators and, finally, systemic anti-infectives. Local products dominate the market. Their share rose from about 55 percent in early 1999 to 65 percent in late 2003. The share of JV-produced drugs and drug imports declined accordingly to about 16 to 18 percent each.

9.5 Conclusion

Novartis Pharma aims to become the leading pharmaceutical company in China by continuing to expand its field force from tier 1 into tier 2 and 3 cities and by maintaining a CAGR of 20 percent or more. This will require it to continue launching innovative products for the Chinese market and making the necessary investments to support these initiatives. Apart from these business-driven results, Novartis Pharma also aims to become the most attractive company to its stake-holders, employees and business partners in China's pharmaceutical market. Provided that the strategies described can be implemented as planned with superior management in place, Novartis Pharma is optimistic of attaining these goals before the end of the decade.

10 China's Approach to Innovative Pharmaceutical R&D: A Review

David E. Webber: WSMI, Ferney-Voltaire, France

This chapter addresses the question of whether a globally competitive innovative biopharmaceutical R&D industry can be developed in China, with participants and investors including both local and foreign-invested companies. Local companies already have substantial assets and knowledge in generics manufacture and low-cost, increasingly high-quality operations. It is easy to envisage these companies becoming the world's principal suppliers of quality generic products. The question addressed here is whether some of them could develop innovative R&D and evolve into research-based pharmaceutical companies. For international R&D-based pharmaceutical companies, the question is whether economic growth coupled with the shaping of a positive environment will encourage increasing investment into innovative pharmaceutical R&D in China.

Based on desk research and particularly on interviews undertaken in China, a study was conducted which developed a view of the issues, prospects and barriers to China contributing to the global effort in pharmaceutical R&D (*Webber* 2003a).[1] The questions asked included: To what extent are the institutions, structures, policies and human resources in place in China? What are the necessary conditions for China to further develop domestic, private-sector research-based companies? What are the opportunities and hurdles for foreign companies in pharmaceutical R&D in China? This chapter provides a brief summary of the previous work, which should be consulted for further details.

10.1 Chinese Pharmaceutical R&D – Current Status

10.1.1 Generics Companies

The Chinese pharmaceutical market for chemical or small-molecule medicines (i.e. not including herbal medicines or traditional Chinese medicines) is largely a generic market, with a modest (but increasing) share taken by innovative patented medicines. In 2003 there were approximately 6,800 Chinese pharmaceutical companies, of which 5,000 produced medicines (small-molecule generics and biotech products) and the remainder were involved in pharmaceutical-related activities

---

[1] Sponsorship of this work and of the summary here by the R&D-Based Pharmaceutical Association in China (RDPAC) is gratefully acknowledged.

such as packaging and equipment supply.[2] Most of these companies lack the size, knowledge and resources to undertake innovative R&D in the short or medium term, and 97 percent of 1,300 synthetic medicines produced in China are copies.

Manufacturing has traditionally been undertaken by companies according to state specification of requirements and with state funding. Fundamentally, given the dominance of generics, the pharmaceutical industry culture in China is strong in meeting the requirements for copying, rather than being oriented towards a dedicated and systematic search for new molecules. A good illustration of this is that to many Chinese companies, even the very term 'R&D' has a different meaning than elsewhere. R&D is used to refer to the production of additional *generic* products particularly for China – in terms of strengths, dosage forms and even specific compounds.

There is considerable duplication of production and resulting over-capacity means that many domestic state-funded companies are not profitable. Steps have been taken by the Chinese authorities to encourage rationalization – most notably by requiring all companies to comply with the standards of Good Manufacturing Practice (GMP) by June 30, 2004. This was expected to lead to a dramatic reduction in the number of firms and a updated figure cited in October 2004 was 3,600 firms (*Zheng* 2004).

In the longer term at least some of the larger current domestic generics producers such as the 999 Company, Hua Bei and Dong Bei are looking to migrate towards truly innovative R&D and innovation focused on small molecules. The North China Pharmaceutical Group is focusing on four areas: biotech products, small-molecule compound chemistry technology, traditional medicines and natural products screening, and formulation technology. Some Chinese companies are substantial entities. Shanghai Pharmaceutical Group, for example, has around 45,000 employees – the equal of many multinationals.

Support from the Chinese authorities for domestic company improvement is considerable and carefully managed. For example, the Pharmaceutical Industry Institute in Shanghai was established as a State Food and Drug Administration (SFDA)-financed institution tasked with assisting domestic companies to improve their R&D. It is housed in the same institute as the Pharmaceutical Industry Information Center of the State Economic and Trade Commission, which provides information to domestic companies on products, and quality and process information. The SFDA and Provincial Drug Administration agencies frequently arrange well-attended pharmaceutical forums discussing topics in R&D, business development and manufacturing, with presentations by government agencies and overseas experts.

---

[2] By 2000, the Chinese pharmaceutical industry had become the second largest supplier of bulk pharmaceuticals in the world, with output of 33,000 tons of antibiotics, 29,000 tons of vitamin products and 45,000 tons of sulfanilamides, analgesics and antipyretics (IMS CHINA UPDATE, Issue 75, February 2000).

10.1.2 Biotechnology – Research Institutes and Industry

Public research institutes and different forms of national R&D programs make up an important part of overall innovation policies. The rationale of such institutions is to support and complement industry's research by conducting more upstream and public research that would not be sufficiently pursued by industry alone. Developing scientific capacity and capability at the cutting edge of science and technology puts in place a key requirement for innovative product-oriented R&D. There is also substantial indirect benefit in the pool of knowledgeable scientists created, who can move between academia and industry. In such ways, universities and research institutes are important facilitators of pharmaceutical R&D.

China has some 200 research institutes for biotechnology and more than 30 of the 150 key state laboratories are focused on biopharmaceutical-related areas. The overall plan for the pharmaceutical industry makes substantial reference to research institute and R&D investment. The Chinese government is today paying great attention to the development of the biotech industry. Many funds have been set up to finance biotech R&D, including the National Natural Science Fund, the Torch Programme, the "863" High-Tech Program and the Five-Year Plans.

The Chinese government has made a concerted drive to move the country into the vanguard of genomics research. In early 1998, the Ministry of Science and Technology established the Chinese National Human Genome Centre based in Beijing and Shanghai, and the Beijing Institute of Genomics, as centers of excellence for genome sequencing and analysis. The establishment of these facilities enabled China to join the International Human Genome Sequencing Consortium in 1999, in which China played a significant part.

China already stands at international level in some areas of research, for example in gene mapping, transgenic technology for animals and plants, gene therapy technology, stem cell research, gene chips and gene research of some major diseases. The country has a number of world-class scientific biomedical institutions - the North and South Genome Centres, The Institute of Materia Medica, Tsinghua and Beijing Universities, for example. However, the *industrialization* of biotechnology still lags behind the Western world, with relatively few biotech companies in existence. Most of the biotech products currently manufactured by Chinese domestic companies are in effect generics – copied biotech molecules produced by molecular cloning and fermentation processes.

Moreover, public research institutions used to be completely isolated from the market. Each institute concentrated solely on its own research designated by the government, resulting in repetition in research, some lack of originality and little focus on commercial opportunities. This leaves a legacy of variable quality research, poor market orientation and lack of experience in the commercialization of research.

10.1.3 Traditional Chinese Medicines

Traditional Chinese medicines (TCM) have played an important role in health care in Chinese and other oriental cultures for thousands of years. In 2001 it was reported that there were 1,036 TCM manufacturers in 31 provinces, although many Chinese pharmaceutical companies include a few TCM products in their portfolios. There are perhaps 4,000 to 5,000 different TCM drug preparations, although if different dosage forms are counted the number would be much higher.

Government policy promotes TCMs and a number of government departments are guiding the TCM industry's attempts to modernize. The Chinese authorities have identified two tracks for developing TCM R&D. The first is through purification and standardization to meet global standards and remove impurities such as pesticides and heavy metals. The second track is to utilize TCM as a starting point for producing *novel* medicines. This may be through the identification and purification of the active element(s) (often complex molecules) or the discovery and development of small molecules which mimic the activity of the original TM. It has been claimed that about 140 new drugs have originated directly or indirectly from Chinese medicinal plants by means of modern scientific methods (*Chang-Xiao Liu/Pei-Gen Xiao* 2002).

10.2 Government Agencies Driving Biopharmaceutical R&D

There are a number of Chinese government agencies that directly and indirectly impact on research and the research-based industry. The most important of these are:

The State Economic and Trade Commission (SETC)

The SETC was responsible for planning the development of the Chinese pharmaceutical industry at a macro level. The 10th five-year plan for the industry was issued in June 2001 and described in some detail the intended direction towards liberalization. The SETC was the key agency with respect to the encouragement of a research-based industry and its focus was on a number of commendable principles, including:

- Encouragement and support of basic research;
- Domestic generics industry quality and rationalization;
- Development of biopharmaceutical technologies and industry;
- Modernization of traditional Chinese medicines.

The SETC has now been disbanded, with its departments being divided and merged into the National Development and Reform Commission (NDRC) and the Ministry of Commerce (MOFCOM).

The Ministry of Science and Technology (MOST)

The MOST is responsible for the overall management of science and technology developments. This covers studying and formulating policies, laws and regulations; evaluating, issuing and organizing implementation of projects; formulating annual programmes; determining the direction of investment; monitoring and checking the implementation of plans; and organizing international cooperation.

The State Food and Drug Administration (SFDA)

The SFDA is responsible for many aspects of drug registration and regulation. Provincial Food and Drug Administrations (PFDA) are central to the implementation of many of these policies.

The National Development and Reform Commission (NDRC)

Formerly known as the State Development Planning Commission (SDPC), the National Development and Reform Commission (NDRC) has a high-technology division and is the agency responsible for setting the prices of medicines. The principal impact on R&D is achieved through supporting pricing of products that reflects the high levels of investment needed for R&D. Similarly the Ministry of Health (MOH), which is responsible for hospital management generally, can affect industry through its policies on hospital bidding and procurement. The Ministry of Labor and Social Security (MOLSS) controls national social insurance scheme drug lists and insurance reimbursement of the patient.

The Ministry of Commerce (MOFCOM)

Formerly the Ministry of Foreign Trade and Economic Cooperation (MOFTEC) and now renamed the Ministry of Commerce (MOFCOM), this body is responsible for investment and business policy generally, for all companies including both domestic and foreign-invested enterprises.

This ministry sets policies for foreign investment, which influences investment decisions on the part of foreign companies. As a result of World Trade Organisation membership, evolution is toward equal treatment for all companies, which must be positive for the longer term.

China's Strategic Orientation in Pharmaceutical R&D

The impression gained in China is that, for understandable reasons of history, the strategic orientation is on generics first and innovation second. Further, within innovation, the focus appears to be on biotech products and TCM, rather than on small molecules.

However, this does not match closely with the global market opportunity and going forwards this may not be the best, or at least a sufficient, emphasis for China. Generics manufacture is a low-margin business and with improved patent protection the opportunity for generics companies to use new technologies free of charge will diminish. As countries become richer, it is likely that the generics sector will lose share going forwards, at least in terms of market value. (This is not to suggest that volumes will not increase substantially, and should do so, to serve the millions of people in the world who do not have access to any medicines. But it will be a decreasingly attractive business.)

Secondly, the growth of the biotechnology industry globally obscures the fact that many companies have yet to show a profit. Only a small proportion of biotechnology companies have succeeded in becoming independent, integrated pharmaceutical companies. Biotechnology products also tend to be limited by the mode of administration (often requiring injection), may be more difficult to manufacture, may be most applicable in a narrow disease situation, have stability problems and can be very expensive. These factors may limit their usage to niche situations. Furthermore, as pharmaceutical science advances, it is possible to envisage situations where small molecules are found that operate on the same targets as biotechnology products but without the problems described. It is no coincidence that the focus of the larger multinational pharmaceutical companies is on small molecules. Thus again it is possible to paint a negative scenario for China's R&D orientation going forwards.

Similar questions can be raised for TCM and herbal medicines generally. TCM usage, even within China, has been challenged by increasing use of 'Western medicines'. The arrival of innovative biotech products and small-molecule NCEs for previously intractable chronic diseases will hasten the process. Challenges to TCM evidence on safety and efficacy can only increase, possibly along with issues of over-harvesting and sustainability (*WHO* 2002). For these reasons TCM may continue to be a popular niche for many years but the value share is unlikely to increase substantially, either in China or in world markets.

One of the conclusions drawn therefore is that the relative emphasis on and support for these sectors deserves fuller discussion. The concern is that the Chinese orientation towards domestic generics companies, biotech and TCM will leave too small an effort directed towards innovative small-molecule NCEs.

10.3 Factors Important in Country Competitiveness in Pharmaceutical R&D

Although there is no universal 'magic formula' that will secure industrial R&D for a country, there are nevertheless a number of factors that collectively determine a country's prospects for developing and retaining R&D. These come under three broad dimensions:

1. **National goals and objectives** of the government and policy makers in a country.
2. **Country structures** – the structures, policies and processes created by government action.

3. **Country resources**, including the human resources and other assets of the country in both the public and private sectors.

National goals and objectives are set by governments, politicians and policy makers and may be explicit and transparent, or implicit and obscure. Country structures cover the legislative and practiced norms of a country, financial structures and essential services. Country structures are of course largely the result of years of government legislation, policy making and the economic conditions. Country resources cover the less structural elements – human capital, skills and knowledge, the natural assets and private industry of a country. Of course, these are also heavily influenced by government action but are distinct in a number of ways – for example they may be more mobile and might even be able to leave a particular country.

These three interlocking dimensions represent a basic framework for assessing the prospects for innovative biopharmaceutical industry at an individual country level. The objective is to consider all the factors necessary to the research-based biopharmaceutical sector. In a previous study, a total of 23 principal country factors were identified as being important in encouraging the development of a research-based pharmaceutical company (*Webber* 2003b). In *Table 10.1* these factors are listed and weighted in terms of their significance in the Chinese situation specifically (*Webber* 2003a).

Table 10.1. The principal factors important to biopharmaceutical R&D in China

Factor impor-tance	1. National goals and objectives	2. Country structures	3. Country resources
Absolutely essential	• Government prioritization	• Intellectual property protection • Government purchasing and pricing policies	• Existing industry • Human resources • Public research/ institutions
Very important	• Government policy coordination	• Trade policies • Regulation • Essential services • Company funding • Tax/fiscal incentives • Domestic market attractiveness	• Education system • Incubators • Natural resources • Related & support-ing industries
Some importance	• Public consent	• Patient need • Business control • Judicial systems • Macroeconomic stability	• Information net-works

Within each of the three dimensions, there are factors that are without question absolute requirements for China, without which an R&D industry could not be successful. These are government prioritization, the existing industry, the research base, intellectual property protection, human resources and government purchasing and pricing policies. The first three of these have already been touched on; brief mention of the other factors follows.

10.3.1 Intellectual Property Protection

China has moved strongly towards protection of intellectual property with reform of patent laws in 1993, and accession to the WTO in 2002 with the attendant commitment to the Trade-Related Intellectual Property Scheme (TRIPS) for the protection of intellectual property. However, China must not rest on the achievements already made in intellectual property protection. Rather, the country needs to maintain momentum by strongly implementing intellectual property rights that are important to the pharmaceutical industry such as data exclusivity and patent linkage. Moreover, China will want to add to its "bundle" or "basket" of intellectual property rights through patent term extension and other exclusivity measures, as in more developed countries, to spur investment into drug development and research. Confidence in how IP policies and processes are enforced going forwards is important to R&D by domestic companies and multinationals alike.

10.3.2 Human Resources

Innovation in pharmaceutical R&D comes from exceptional individuals plus teams of people – scientists, technologists, engineers and technicians, working together with administrators and managers. China has many science graduates and there is strong competition for laboratory jobs. There are some 200,000 researchers specialized in biotech R&D in China and a previous 'brain drain' of scientists is being reversed, with many returning home. A web of personal contacts, or *guanxi*, with Chinese scientists working overseas also helps facilitate the transfer of ideas, personnel and funding back to China. However, these positive statistics for human resources are somewhat misleading as they relate mainly to graduates, whereas there is a chronic shortage of postgraduate and PhD-level personnel, entrepreneurs and business managers familiar both with China and with the practices of the market economy.

10.3.3 Government Purchasing and Pricing Policies

Most worryingly, the dominant role of the state in China may inadvertently slow the development of a competitive research-based industry. In the Western model of industrial R&D, governments invest in 'pure' or 'basic' research. The private sector undertakes some basic research and accesses the pool of public research in the search for targets and potential products. Companies are responsible for the process of finding and bringing products to the market, which is time-consuming and extremely expensive. The device that is used to finance companies for investing in R&D and bringing products to market is the patent system. Protected by patents companies can sell products at prices that recoup the cost of past and current R&D and incentivize further research.

The patent system component may be in place but the Chinese approach appears to be limiting, particularly with regard to pricing incentives. State-financed research institutes link to state-funded generics companies that sell products at state-mandated 'generic' prices. These prices are cost-based, to which is added a small percentage for profit. This is a 'push' model for new product development. However, nowhere in the world has a push model led to a viable, competitive industry.[3] By contrast, the global pharmaceutical industry is made up of independent, self-funded or risk capital funded companies and is driven by 'pull' from markets that provide an adequate return for R&D.

This has important implications for China. In implementing intellectual property protection, the Chinese government has put in place a key part of what is necessary to reward investment and innovation in biopharmaceutical R&D. At the same time, *patent protection only has value if prices reflect the levels and risk of R&D investment, rather than the cost of product manufacture.* This is the core principle that should underpin the multiple strands of every government's policy on the financing of innovative drugs, the pricing of all drugs, the creation of an indigenous pharmaceutical industry and encouragement of foreign investment. Chinese research-based industry (domestic and foreign) will develop in direct proportion to the premium that is allowed by the systems that determine the reimbursement, pricing and purchase of innovative, patented medicines. Chinese R&D is likely to have a very slow lift-off if it relies principally on state push support, low domestic market pull and sales in other countries.

10.4 Collaboration with the Multinational Pharmaceutical Companies

Little reference is made in Chinese government strategy statements to the contribution that multinationals might make, or to incentivising their involvement in technology transfer or local R&D. The view appears to be that multinationals should simply provide support to the global marketing efforts of Chinese companies or fund R&D laboratories in China. This is a narrow orientation and not indicative of the real collaborative opportunities. From a multinational perspective, the first phase of engagement in China – through manufacturing – is maturing. Under the right circumstances, a new phase involving significant R&D collaboration in certain areas could now emerge.

---

[3] See *Webber* 2003b for a discussion of R&D models. The state-managed approach also suffers from problems with skills, knowledge & facilities, goal orientation and financing.

However, the multinational companies are unlikely to parachute large research facilities into China in the short or medium term as the conditions and incentives are currently insufficient.[4] But neither should multinational companies (or foreign 'biotech' companies) be simply viewed as competitors to domestic companies - the opportunities to develop strategic relationships between the different parties are considerable and multinational companies are likely to be very supportive in this regard.

For 'big pharma', potential collaborators in R&D can have input at many points along the value chain, adding value and reducing risk. Anything that can impact positively on any part of the value chain is valuable. Benefits to pharmaceutical companies include increased access to scientific centers of excellence, the leverage of funding and science, access to developing technology and new skills, potential access to targets or even products for development and the opportunity to access potential recruits.

Benefits for Chinese biotechs and academia from collaboration with the multinationals include access to funding (including improved prospects for accessing venture capital), intellectual scientific direction, state-of-the-art platform technology and facilities and the translation of basic research into usable products.

10.4.1 Research

The multinational approach to R&D is through target identification and optimization, ultra-high-throughput screening using combinatorial chemistry approaches, and small-molecule identification and optimization. Developments in science and technology and fragmentation and *de facto* outsourcing of individual components of these processes are offering increasing scope for new entrants to gain a foothold in this R&D chain through specialization. There are niches for focused, nimble, technology-based companies concentrating on adding value through a disease, product or technology focus. For companies with limited R&D budgets, strategies for entry at particular points or specializations can reduce risk and greatly reduce the entry costs. This represents an avenue for entry for Chinese companies and researchers – through products, services and research tools.

The biotech approach is important for innovation not only towards biotech products but also towards *tools* for small-molecule development. Biotechnology tools today underpin a great deal of the new research approaches and many companies offering technology services and platforms have been created to serve this need.

Universities and biotechs, like pharmaceutical companies, operate in global markets and need to identify niche areas where they are truly world-class. As the institutions and companies that are able to conduct innovative research emerge,

---

[4] Contrast with the Singapore Economic Development Board's S\$ 1 billion (US\$ 550 million) R&D fund set up in June 2000 to attract foreign companies. The Research Incentive Scheme for Companies in Singapore provides *up to 50 percent of the total research expenditure* of a company over five years.

they will attract support from technology-driven companies, including the multi-national companies. This is happening already to some degree but many opportunities for research collaborations remain.

The incremental strategies for foreign companies to participate in and build experience in pharmaceutical research in China could also include collaborative research and technology-related consulting, licensing in and out of technologies, patent exchanges, personnel exchanges, cooperation in the education of graduate students, vocational training for employees, assistance with start-ups and informal contacts and personal networks.

10.4.2 Development – Clinical Trials

Further down the value chain, there is today a specific opportunity to develop China as a leading country for international clinical trials. The global market for clinical trial work is highly competitive, but there is a shortage of patients and good researchers in some disease areas. In general, as the cost of clinical trials continues to soar, the opportunity to use developing country sites (usually with lower doctor, scientist and patient recruitment costs) becomes ever more attractive.

The benefits for China and for companies (domestic and multinational) are clear. Participation by leading hospital centers and research institutes in global clinical trials provides the opportunity to work on cutting-edge medicines. Spin-off benefits include technology transfer and training in Good Clinical Practice, an important part of the R&D value chain where China has been weak. Funding and other resources used for trials not only provide welcome income but can be used in the development of the institutions involved.

This will provide a basis for expansion up and down the value chain. Down the value chain, the product patent holder may be better positioned to work with local doctors and companies to market the product. Up the value chain, if the leading doctors and scientists are involved with the latest technologies, it encourages companies to undertake future trials and earlier stage work in the country concerned.

However, a number of areas need to be addressed, most notably the time taken for regulatory approval for proposed trials, a lack of centers and trained investigators and various procedural issues.

10.5 Conclusion

In conclusion, China has in place many of the factors necessary for innovative pharmaceutical R&D. The country has a strategic focus on this area, a large existing industry, some strong public research institutions and other assets. Some factors of importance such as intellectual property protection and government purchasing and pricing policies need ongoing attention. The opportunity exists today for multinational companies to move on from manufacturing-led investments in China to collaborate in various ways in R&D. This is already happening but the potential for further development is great. The benefits both to the multinational companies and to Chinese institutions and companies are clear.

In summary, with accurate navigation and assuming the commitment towards a market economy is resolute and carried through to completion, China could expect in future to join the select group of countries contributing to innovative biopharmaceutical R&D.

11 Foreign Direct Investment by Multinational Corporations in China – The Pharmaceutical Sector

Xiangdong Chen: Beijing University of Aeronautics & Astronautics, P.R. China;
Guido Reger: University of Potsdam, Germany

11.1 Introduction

The pharmaceutical industry and its innovation process are highly international-ized. This is true for the exploitation of innovations generated in the home coun-try (export of new drugs), production facilities in foreign countries, the applica-tion of a new drug to specific regulations in national markets, cross-border technology-related collaborations or alliances, and the generation of innovation in countries outside the home base (*Jungmittag/Reger/Reiss* 2000). Our research here focuses on foreign direct investment (FDI) by multinational pharmaceutical companies in China.

We used different sources for this investigation, among them secondary statis-tical data from pharmaceutical and enterprise associations and the Chinese Min-istry of Commerce, patents submitted to the State Intellectual Property Office of PRC and a database provided by the Delegation of German Industry and Com-merce in Shanghai. However, our task was not easy and, all in all, our data search shows that at least some of the data are incomplete and far from being rep-resentative. Obviously, there is still some research regarding data and indicators to be done before the puzzle can be completed and give a whole picture.

This contribution is structured as follows: section 11.2 describes the trends and structures of foreign direct investment in China. The relationship between tech-nology, patents and foreign direct investment by multinational companies in the Chinese pharmaceuticals market are analyzed in section 11.3. Conclusions are drawn in section 11.4.

11.2 Foreign Direct Investment (FDI) in the Pharmaceutical Sector in China: Trends and Structures

11.2.1 Trends in FDI in the Pharmaceutical Sector in China

Foreign direct investment (FDI) in China has been developed continuously over the past decades and in 2002, the volume of overseas investment into China ranked first in the world. In recent years, the main sectors of foreign direct investment have ranged from manufacturing through services, and among these, the pharmaceutical sector is one of the most promising areas.

The pharmaceutical industry in China, along with related fields such as medical instrument and equipment manufacture, continues to expand and the market is attracting investment and business input both from domestic and overseas companies.

On average, the annual increase in production value of Chinese pharmaceutical companies has been around 19.5 percent since 1991. In 2001, the production value of China's pharmaceutical industry increased to RMB 265 billion, equivalent to US\$ 32.1 billion (*Chen* 2003). It is estimated that the market for medicines in China will expand by 15 to 20 percent annually over the next five years and may be as high as US\$ 120 billion by 2020 (*Liao* 2003). If this becomes reality, the pharmaceutical market in China will emerge as the world's largest single pharmaceuticals market.

This significant growth can be explained by two reasons: the increasing needs of China's vast population and international medical production transfer. Cities in China are changing significantly in terms of economic conditions and local living standards. Therefore, diversified demand for medical and healthcare products and services has been greatly emphasized in recent years, which has also meant an expansion of the market for higher value-added medicines. On the other hand, the pharmaceutical industry covers a wide range of production in terms of technology advances and raw materials. As Vernon's theory (*Vernon* 1966, *Vernon/Wells* 1986) on product life cycle and internationalization explains, matured technology and production are generally transferred from highly developed countries to developing countries and regions, especially in those sectors such as medicine production which are often characterized by higher environment control and raw material costs.

However, the capacity of the local Chinese pharmaceutical industry is still low compared with the international level. Although China exports increasingly large quantities of medical products to the world market, most of them are raw materials. In the early 2000s, exports amounted to around US\$ 2 to 3 billion and included medical preparations worth only around US\$ 100 million. According to an international exhibition of raw materials for pharmaceutical products held in Shanghai in 2004, China now holds second place for raw material production and export to the world pharmaceutical market.

Compared with local firms, overseas pharmaceutical companies are much more advanced in their downstream products and more aggressive in their market promotion. The top multinational pharmaceutical company, Pfizer along, for example, generate annual revenue more than 50% of total China's medical market sales in 2004. The results are often significant and profound. A survey conducted in China shows that among the 50 most favorite medicines, 40 are produced overseas and imported (*Special Report* 2004a).

Ultimately, local manufacturers still have a weak technology base for new product design and production compared with international companies. Of the some 3,000 manufactured medicines produced in China, 99 percent are imitations of products from companies abroad. In recent years, of 837 newly developed medicines, 97 percent are imitations (see www.chinapharm.com.cn, December 2002) and about 60 percent are still covered by patent license arrangements. Nonpatent medicine production is considered very promising in the near future in China, due to the expiration of more than 150 important patents (*News Report* 2003).

The large and rapidly expanding market with a lower local production level makes it possible for overseas pharmaceutical companies – especially large multinational pharmaceutical corporations – to enter the Chinese market more easily and vigorously. In fact, the pharmaceutical market in China was opened to Western companies quite early in the 1980s. The first joint venture in pharmaceuticals was Otsuka established in Shanghai in 1980. Xi'an-Janssen, Tianjin Smith Kline and others were all pioneers in exploring the opportunities offered by the Chinese market. From that time on, large international pharmaceutical companies focused increasingly on the Chinese market and investment in this area grew to US$ 150 million. It is reported that the number of foreign companies investing in the medical sectors has already increased to around 1,790 with a realized investment volume of more than US$ 1.5 billion. Of these companies, 645 are active in the pharmaceutical sector. Among the largest 25 multinational pharmaceutical corporations in the world, 20 are currently operating in China (*Guo* 2003). 40 percent of Chinese local medical firms include some degree of foreign direct investment.

Table 11.1 shows the shares of FDI firms on various measures to overall industries, which suggest that overseas investment plays an important role in the Chinese pharmaceutical industry. According to another study by the State Economic & Trade Commission, 645 foreign pharmaceutical companies (accounting for only 17.5 percent of the firms in the total sample) contribute 23.6 percent of annual sales and generate 25.8 percent of the profit in the industry (*Special Report* 2004b). Based on *Table 11.1*, it is clear that FDI medical firms generate major part of profits and production value as well as export value in medical packaging, medical equipment, pharmacy equipment, and sanitary materials sectors. However, in terms of absolute volume of production, revenue, and profit, foreign medical companies performed better in chemical medicine, medical equipment, and Chinese traditional medicine. On the other hand, FDI firms in all medical sectors are very much unevenly distributed. *Fig. 11.1* illustrates the distribution of FDI within the medical sector. Chemical medicine, medical equipment and traditional Chinese medicine are the three largest areas of foreign direct investment.

Table 11.1. Shares of FDI firms in pharmaceuticals in China (million RMB, %) (Source: based on data from China Medical Economic Research Centre, www.chinapharm.com.cn)

Sub-sectors	Shares of FDI firms in the sector	Product value / export		Revenues		Profits in the sector	
		Share of the sector	Share by FDI firms	Share of the sector	Share by FDI firms	Share of the sector	Share by FDI firms
All medical sectors	17.5 %	100 %	28.4 %	100 %	23.6 %	100 %	25.8 %
Chemical medicine	18.1 %	67.45 %	12.6 %	60.09 %	21.4 %	55.67 %	23.4 %
Medical equipment	20.7 %	17.49 %	68.5 %	7.52 %	39.6 %	5.70 %	39.3 %
Sanitation materials	24.6 %	7.90 %	59.6 %	1.91 %	43.8 %	0.70 %	48.8 %
Pharmacy equipment	10.0 %	0.13 %	27.8 %	0.35 %	12.6 %	0.17 %	41.3 %
Traditional Chinese medicine	15.2 %	4.09 %	27.0 %	25.76 %	22.5 %	33.99 %	23.0 %
Troche (Chinese medicine)	20.7 %	0.37 %	96.8 %	0.46 %	15.7 %	0.24 %	30.0 %
Medicine packaging	10.9 %	1.18 %	77.9 %	2.48 %	23.3 %	1.65 %	45.4 %

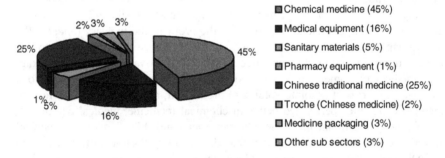

Fig. 11.1. Distribution of foreign invested firms (by numbers of firms) in pharmaceutical related industries in China (Source: based on data from China Pharmaceuticals, see www.chinapharm.com.cn)

11.2.2 FDI in Pharmaceuticals Compared with Other Manufacturing Sectors

Currently, there are about 5,000 firms operating actively in the pharmaceutical sector in China. The local pharmaceutical industry in China is a small sector in terms of production value, total fixed assets and market revenue. It generates about 2 percent of overall industrial production nationwide, ranking 20th among 39 different industries in China. However, the contribution of this sector is significantly high compared with others. For example, based on its 1.93 percent share of all industrial assets, the pharmaceutical sector contributes around 2.6 percent of taxes and 4.27 percent of total manufacturing industry profit in China. The pharmaceutical sector ranks 7th among the 39 industries in terms of profit earnings. At the same time, it is estimated that foreign firms can achieve a profit rate of 20 percent or more.

Fig. 11.2 illustrates the level of FDI in the pharmaceutical sector compared with other branches of industry in China. The data show that although foreign direct investment in pharmaceuticals in China is not high in absolute terms, the investment projects are fairly large on average compared with other branches and manufacturing industry in general. Foreign firms in the pharmaceuticals sector have large-sized investment projects compared with overseas companies in other branches of manufacturing, especially in the categories Large II, Mid I and Mid II (see *Table 11.2*).

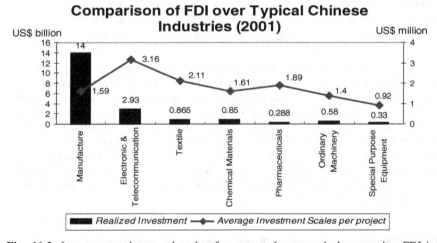

Fig. 11.2. Investment volume and scale of overseas pharmaceutical companies: FDI in pharmaceuticals compared with other industries in China 2001. (Source: based on data from the Chinese Ministry of Commerce, see www.mofcom.gov.cn)

Other interesting characteristics of FDI in pharmaceuticals in China are worth mention. One is that overseas companies prefer to use joint ventures rather than wholly owned firms as their investment instrument (only 23.5 percent of investment in the pharmaceutical industry is represented by wholly owned firms). This difference is significant if one compares it with FDI in such areas as electronics and telecommunication where 60 percent of investment is represented by wholly owned firms. This may be reasonable as pharmaceuticals usually involve multiple local players and marketing channels. Another difference is that companies from Hong Kong, Taiwan and Macao play a minor role in the pharmaceutical sector in mainland China (48 percent of pharmaceuticals, referring to the average level of 60 percent in total manufacturing sectors, and the highest level of 71 percent in textiles).

Table 11.2. Comparison of pharmaceutical companies' investment in China according to size

Size	Size distribution of all FDI pharmaceutical firms (left) compared with FDI firms in all sectors (right)		Size distribution among major investment modes by pharmaceutical companies from developed countries (excluding investment from Hong Kong, Taiwan, Macao)	
			Joint venture	Wholly owned
S. Large	0.4%	0.24%	0.4%	1.9%
Large I	1.9%	1.7%	3.2%	1.9%
Large II	3.7%	2.3%	4.0%	2.6%
Md I	3.9%	2.5%	3.6%	3.8%
Mid II	6.8%	3.7%	6.5%	1.9%
Small	83.3%	89.5%	82.2%	87.8%
Total	100%	100%	100%	100%

Note: According to the Chinese Statistical Bureau, industrial firms are divided into six categories in terms of the value of their fixed assets and the number of employees. The category named Special Large Firms includes those companies with a value of fixed assets (VFA) greater than RMB 0.5 billion, equivalent to US$ 60.5 million; Large I type: between RMB 0.3 billion and 0.5 billion (US$ 36.3 million to 60.5 million); Large II: between RMB 0.1 billion and 0.3 billion (US$ 12.1 million to 36.3 million); Mid II: between RMB 0.05 billion and 0.1 billion (US$ 6.1 million to 12.1 million); Mid I: between RMB 0.01 billion and 0.05 billion (US$ 1.2 million to 6.1 million); Small: less than RMB 10 million (less than US$ 1.2 million). (Source: edited from Chinese enterprise website, see www.ceie.com.cn)

11.2.3 Structure of Foreign Direct Investment by German Chemical and Pharmaceutical Firms in China

This section is based on an analysis of a database which has been provided by the Delegation of German Industry and Commerce (GIC) in Shanghai (hereafter referred to as the GIC database). This database includes a total of 1,515 records of German firms that had invested in China by January 2001. The data are provided by the firms on a voluntary basis and are not complete. In particular, sensitive data about the size of the investment, employees or annual sales in China are not included. Moreover, the GIC database was not built for scientific purposes and is not statistically representative of German FDI in China. However, due to the lack of other more comprehensive data sources, it gives a first impression of the structure of investment by German firms in China.

For our analysis, German firms in the chemical and pharmaceutical sector were selected from the GIC database. Since the assignment of an individual company to a sub-branch is not clear-cut in each case, it was not possible to perform the analysis for pharmaceuticals alone. According to the definition of the database, this sector includes the sub-branches outlined in *Table 11.3*. Of the 1,515 German firms, 205 (13.5 percent) are active in the chemical and pharmaceutical sector.

Table 11.3. Sub-branches of the chemical and pharmaceutical sector (Source: GIC database)

Code	Sub-branch	Code	Sub-branch
B0701	Chemical Products, General	B0708	Paints and Varnishes (Pigments)
B0702	Inorganic Elements (Gases)	B0709	Glues, Adhesives, Putties, Cements
B0703	Organic Elements (Rubber)	B0710	Chemical for Agriculture (Fertilizers, Insecticides)
B0704	Essential Oils, Essences, Aromas	B0711	Plastics, Semi-Finished Plastic Products, Fibers (Plastics for Medical Use)
B0705	Pharmaceutical Products	B0712	Lubricants
B0706	Cosmetics	B0713	Anti-corrosive Materials
B0707	Wax Products, Polishes and Cleaners (Detergents)	B0714	Chemical Materials

We analyzed the structure of FDI by the German companies according to the three dimensions: (1) foundation date, (2) type of FDI and (3) foreign-owned stake in the equity joint ventures.

The year of founding a subsidiary or firm by German chemical and pharmaceutical companies in China covers the period from 1980 to 2000 in the GIC database. 23 of the 205 German firms in the sector did not provide the information so this part of the analysis builds on 182 answers.

The quantity of investment in China dramatically increased between 1993 and the end of the 1990s (see *Fig. 11.3*). 152 (83.5 percent) investments can be counted between 1993 and 2000, with the highest numbers in 1993 and 1994. In the 1980s the number of investments was quite low. Between 1980 and 1992 only 16.5 percent of investments in this sector were made (absolute number: 30). After the boom period in the mid-1990s the number of investments slowed at the end of the decade. The large drop in the number of investing firms after 1998 could be explained as a negative impact of the Asian financial crisis. Furthermore, the low figures in 1999 and 2000 could be distorted by a bias in the GIC database because of a lack of responses from the companies.

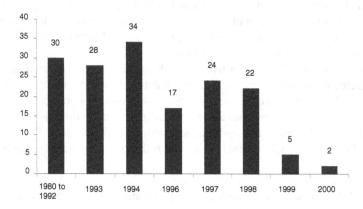

Fig. 11.3. Year of foundation by German chemical and pharmaceutical firms in China (Source: GIC database)

According to the GIC database, the modes of investment were divided into five possible forms: representative office, contractual joint venture, equity joint venture, wholly foreign-owned enterprises, and branch company. Equity joint ventures clearly dominate the types of investment with 45.4 percent (absolute number: 93) followed by representative offices with 31.7 percent (absolute number: 65) (see *Fig. 11.4*). Both forms account for more than three quarters of the total numbers of types of investment. Wholly foreign-owned enterprises (14.1 percent; absolute number: 29), branch companies (4.4 percent; absolute number: 9) and contractual joint ventures (4.4 percent; absolute number: 9) play virtually no role as an entry mode into the Chinese market.

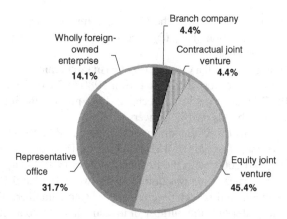

Fig. 11.4. Mode of investment of German chemical and pharmaceutical companies in China (Source: GIC database)

Whereas the other types of investment are usually wholly owned by the German firm, the question of ownership and foreign-owned stakes is of interest regarding the equity joint ventures. 80 of the 93 equity joint ventures provided information about their share of the equity. Majority ownership plays the most prominent role in the equity investment in China by the German chemical and pharmaceutical firms questioned. 56.3 percent (absolute number: 45) of the equity joint ventures are owned to between 51 and 80 percent by the German firm; 13.8 percent (absolute number: 11) are owned to between 81 and 99 percent by the German company (see *Fig. 11.5*). Together these amount to more than 70 percent. In contrast, minority ownership plays a small role: 16.2 percent of the equity joint ventures have a German-owned stake of 25 to 49 percent and 13.8 percent have a German-owned stake of 50 percent.

The high relevance of majority ownership points to the interest of German chemical and pharmaceutical corporations in controlling their activities in China and managing their investment in German style. It also indicates that German corporations consider their investments in China to be of high importance and of a long-term nature.

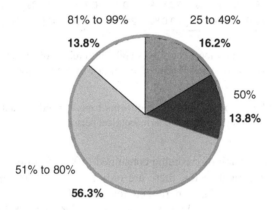

Fig. 11.5. Stakes held by German chemical and pharmaceutical corporations in equity joint ventures in China (Source: GIC database)

11.2.4 Multinational Pharmaceutical Companies from Selected Countries Investing in China

Multinational pharmaceutical companies have been focusing their business ever since mid 1990's. In recent years, top 20 multinational pharmaceutical corporations have all invested in China. With the merger of Glaxo Wellcome and Smith Kline Beecham, Glaxo Smith Kline (GSK) founded the largest pharmaceutical company in China. With its three-percent market share, the company heads the list of all foreign pharmaceutical firms in China. GSK even ambitiously aims at an annual market revenue of GBP 250 million and expanding its market share to 7.3 percent (see www.sinopharm.com). *Table 11.4* shows the distribution of foreign firms from nine countries which represent 71.5 percent of the total FDI firms in the field. It is interesting to note that all these countries also represent the leading exporters to the Chinese pharmaceutical market. For example, the largest volume (21 percent) of pharmaceutical imports to China is from the United States. 14 percent come from Japan, 13 percent from Germany and 24 percent from other European countries including France, Italy, the United Kingdom, Switzerland and Belgium.

Table 11.4. FDI in the Chinese pharmaceutical industry: country of origin and size (Source: based on the Chinese Enterprise Product Database, see www.ceie.com.cn)

	United States	Japan	Korea	United Kingdom	Ger-many	France	Switzer-land	Nether-lands	Sweden
Dominant share*	57.0 %	16.1 %	7.6 %	6.4 %	4.7 %	2.8 %	1.8 %	1.2 %	0.8 %
Large	8.3 %	18.8 %	10.0 %	5.9 %	24.0 %	93.3 %	38.5 %	22.2 %	33.3 %
Mid	9.3 %	8.2 %	2.5 %	23.5 %	4.0 %	6.7 %	23.1 %	11.1 %	16.7 %
Small	82.4 %	72.9 %	87.5 %	70.6 %	72.0 %	0.0 %	38.5 %	66.7 %	50.0 %

Note: This database shows 3,068 local companies and 738 foreign companies in the pharmaceutical sector. The company size is classified as defined by the Chinese Statistical Bureau (see categories in *Table 11.2*).

*Dominant share is indicated by the percentage of firms from a specific country to the total number of foreign firms in the Chinese pharmaceutical sector.

If measured by the volume of medicine consumed (2001) from the top 20 producers in the pharmaceutical sector, there are only four local companies. Measured by annual sales, Xi'an-Janssen ranked top, with Livzon, Sino-US Smith Kline and Shanghai Squibb all in the top ten (see *Special Report* 2004b).

In order to keep their competitive advantage, several large multinational pharmaceutical corporations have very actively improved their technology potential. In recent years, R&D investment in China by multinational companies has increased, which is also a promising sign for future investment in the Chinese pharmaceutical sector.

In 2004, Swiss company Roche's fifth R&D center was completed in China and will host 40 to 50 Chinese scientists. Roche ranks fifth behind Novo Nordisk, AstraZeneca, Eli Lilly and Servier as a foreign pharmaceutical company investing in R&D in China. This movement implies that, besides the motives for entering the local market, multinational companies expect other benefits from their investment. It is generally assumed that China may have some advantages in attracting both production facilities and R&D activities. These include low R&D personnel costs, a broad spectrum of disease records resulting from the country's diversified nationalities (156 nationalities live in China), China's unique tradition of plant-based medicine, and pharmaceutical resources such as raw materials.

11.3 Technology, Patents and Foreign Direct Investment in China by Multinational Pharmaceutical Companies

11.3.1 Technology and Foreign Direct Investment by Multinationals in Pharmaceuticals

There are several important features of foreign direct investment and the relationship with technology. First of all, overseas investment in the Chinese pharmaceutical market is closely related to the export of medicines to China. Generally speaking, large and growing markets attract corresponding investment in order to expand the investors' market shares. In fact, the import of overseas medical products to China has grown dramatically in recent years, mainly due to two reasons. Firstly, the domestic market demand for highly effective medicines has increased and secondly, foreign companies have promoted their medical products effectively. High value-added medicines comprise the bulk of imports to the Chinese market. For example, import quantities increased by 4.2 percent during the late 1990s. However, at the same time, import values increased by 114 percent. Today, imported and manufactured pharmaceutical products by foreign firms supply one third of the Chinese market. In large cities, the figure is as high as one half. In some coastal cities such as Guangzhou, the volume of imported medicine is two times higher than that of locally produced ones. In general, these are all higher value-added products. Apparently, pharmaceutical companies and their products are strongly supported by advanced technologies and new scientific findings.

Protecting this scientific and technical knowledge may be one reason why overseas investment by pharmaceutical companies is strictly controlled in terms of ownership. Among the largest 500 joint ventures with foreign equity there are 14 in the pharmaceutical sector, of which 93 percent are controlled through ownership majority. For example, in the case of Xi'an-Janssen, the foreign company holds 52 percent of the equity and in the case of Tianjin Smith Kline the multinational holds 55 percent (*Wang* 2002). For the pharmaceutical joint ventures established in recent years, overseas investors often claim higher stakes up to 90 percent or more. Moreover, patent applications in the area of medicine-related technologies by foreign companies in China are growing faster than the average patent application rate.

In recent years, multinational pharmaceutical companies in China have restructured their production systems through outsourcing and buying back products. For example, Japan's First Pharmaceuticals was the inventor of Ofloxacin.

However, the company is currently buying from China's Zhejiang KangYu Pharmaceuticals. Raw materials for pharmaceutical production seem to be one of the major business areas of local Chinese producers. Today, China has five categories of raw materials that hold top positions worldwide, among them Penicillium which accounts for 60 percent of the world market, vitamin C which accounts for 50 percent of world supplies, and others like Doxycycline HCl and cephalosporin.

11.3.2 Patenting Activities in China: Trends and Main Players

If a market gains in importance, usually patent activities by companies increase because – as a strategic measure to safeguard and enhance market shares – the technology behind products is protected in this specific market by patents. Since the mid-1990s the number of patents in China has increased dramatically, primarily from company applicants. Among these, foreign firms play a major role. It is estimated that the number of patent applications made by multinationals in China has grown by 30 percent annually, especially in high-tech fields like telecommunications, computer technology, home appliances and, above all, pharmaceuticals.

In the Chinese patent system, there are three different types of patents - invention, utility model, industrial design. Of these, invention patents are considered to be technologically the most sophisticated but still far away from market implementation. On the other hand, utility model and industrial design patents are less sophisticated and closer to market implementation. In general, the majority of invention patents are owned by foreign firms or organizations, while the other two types of patents are mostly owned by local companies or applicants from Hong Kong, Taiwan and Macao. Invention patents are therefore the major focus of this section. *Fig. 11.6* illustrates the development of invention patents in the pharmaceuticals sector in China which are mainly owned by foreign multinationals. The patents presented in *Fig. 11.6* cover the four areas of medicine and hygiene, organic chemistry, organic micromolecules and biochemistry as defined by the International Patent Classification IPC. The increase in patenting activities, particularly on inventions, can be taken as an indicator for the strong commitment of pharmaceutical multinationals to the Chinese market and the will to produce in China.

Furthermore, according to the patent database of the State Intellectual Property Office of PRC, the patenting performance of the foreign pharmaceutical companies has been very active in recent years. For example, Pfizer applied to the Chinese State Intellectual Property Office for a total of 768 patents and is the biggest foreign applicant and patent owner in the overall pharmaceutical sector in China. Pfizer focuses its innovations on organic heterocyclic compounds (about 50 percent of all patents), an important area for new medicine design and production. 35 percent of its patents are related to medical and dental surgery and 15 percent are divided between organic compounds, microorganisms, enzymes, and variation & gene engineering. Japanese multinationals like Takeda are also very active here. The company holds 501 invention patents, primarily in the field of organic chemistry.

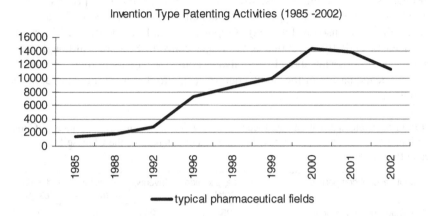

Fig. 11.6. Development of patenting activities in four pharmaceutical areas in China (Source: edited from the Chinese State Intellectual Property Office website, see www.cnipr.com)

Fig. 11.7 presents the patenting activities of foreign organizations (mainly multinationals) in four different pharmaceutical fields in China. The distribution is similar to that of FDI by foreign firms. Organizations from the United States, Japan, Germany and the United Kingdom are at the forefront here.

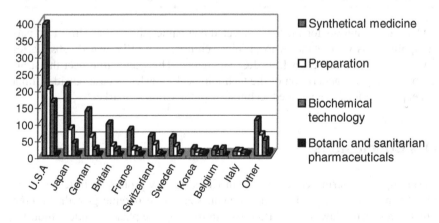

Fig. 11.7. Patenting activities in China in 2001, distribution by country (Source: based on the Chinese Pharmaceuticals Yearbook 2002 – 2003)

The strategic focus of foreign firms can also be revealed by analysis of their patenting activities in China. It is interesting that in the pharmaceutical sector, multinationals can generate and distribute their technologies through numbers of subsidiaries or joint ventures elsewhere in the world. *Table 11.5* presents information on the number and types of patents held by three selected multinationals (Merck, Pfizer, Johnson & Johnson). Keeping the small sample in mind, it can be concluded that, firstly, invention is the major concern. Secondly, although joint ventures or subsidiaries abroad seem also to apply for patents in their host countries, the headquarters in the home country seems to dominate the generation of technology.

Table 11.5. Patent applications by three multinational pharmaceutical companies in China by origin of application (Source: based on the patent database of the Chinese State Intellectual Property Office)

Multinational companies	Patent application total	Invention	Utility model	Industrial design
Merck (from Germany)	254	254	0	0
Merck (from UK)	21	21	0	0
Merck (from US)	12	9	0	3
Pfizer (from US)	715	715	0	0
Pfizer (from UK)	5	5	0	0
Johnson & Johnson (from US)	95	71	0	24
J&J (from Germany)	9	9	0	0
J&J (from Japan)	7	6	0	1
J&J (from Shanghai, China)	30	6	5	19

Our patent analysis for the three selected companies shows that almost no patent applications were made by local joint ventures or wholly owned subsidiaries in China (except a few cases from J&J Shanghai). This seems to reflect the strategy of keeping the generation of the technological knowledge base outside China and exploiting locally generated technology in the Chinese market.

11.4 Conclusions

To sum up, the pharmaceutical sector in China is very promising in terms of its attractiveness for foreign direct investment, size and economic growth. Foreign direct investment plays an important role in the Chinese pharmaceutical industry regarding, both with regards volume and technological expertise. Compared with other manufacturing sectors, most foreign industrial investors in this sector are based in the United States, Western Europe and Japan, with a small number coming from Hong Kong, Taiwan or Macao. Also, the average investment project is fairly large in comparison with other branches.

Our analysis shows that China is increasingly important for multinational pharmaceutical companies. This is true not only as a large and growing potential market, but also as a base for manufacturing advanced medicines. However, in terms of product innovation and the technological knowledge base, until now technologies have been generated outside China. The mode of technology-related investment can nevertheless be characterized by home-base exploitation and not by home-base augmentation. However, due to the obvious strong commitment of the multinational companies and the strategic importance of their investment in China, technology resources can be true cornerstones to improving the competitive advantage of the pharmaceutical industry in China in the future.

With the support of a database, we specifically analyzed the FDI behavior of German companies. They invested heavily in the Chinese market between 1993 and 2000, with peaks in 1993 and 1994, following a quite low amount of investment in the 1980s. After the boom period in the mid-1990s, the number of investments declined towards the end of the 1990s, which may reflect the negative impact of the Asian financial crisis. A further feature is that equity joint ventures and representative offices clearly dominate the various forms of investment. Within equity joint ventures, majority ownership plays the most prominent role, whereas minority ownership is less important. The high relevance of majority ownership indicates that German corporations consider their investment in China to be of high importance and strategic relevance.

Acknowledgement

The authors wish to thank Dr. Gunter Festel and Prof. Maximilian von Zedtwitz for organizing and editing this book on the chemical and pharmaceutical industry in China. Zhang Chen, Jiang Hua-an and Dana Mietzner supported our work by screening various sources, collecting necessary information and analyzing the database; their support is gratefully acknowledged.

12 Competing in the Chinese Antibiotics Market – Cephalosporins 1982-2000

Gail E. Henderson: University of North Carolina, Chapel Hill, USA;
William A. Fischer: IMD, Lausanne, Switzerland

The nature of competition is changing within China's markets, and antibiotics are no exception. This chapter looks at a specific sub-set of the antibiotics market – cephalosporins – and examines how competition has evolved in the wake of China's economic reforms. First-generation cephalosporins were initially introduced into China by Bristol-Myers Squibb in 1982. Prior to this, there was no domestic cephalosporin production in China. In the ensuing 18 years, until the end of the millennium, considerable competition has been engendered in this important segment of the antibiotics market. Today, both Chinese and foreign players compete over all three generations of cephalosporins, across the country, in a manner that tells us much about how competition is unfolding within the Chinese market. This paper traces the evolution of the Chinese cephalosporin market from 1982 until the end of the twentieth century, and offers some hypotheses as to what the implications are for competing in many similar Chinese markets.

12.1 Cephalosporins and China

Cephalosporins are a type of antibacterial agent generated by a fungus. They have the same basic structure as the penicillins (a beta-lactam ring), which inhibits the formation of the protective cell wall of many bacteria. However, because cephalosporins are resistant to the bacterial enzymes that degrade penicillins, they kill a much broader range of bacteria. This broader spectrum of the cephalosporins made them extremely popular in the 1970s, and spawned the focused development efforts of the next decade that produced several generations of semi-synthetic compounds with ever-wider activity (*Cohen* 1984, *Mandell/Douglas/Bennett* 1990).

Cephalosporins have been divided into generations based on the time of their development and biological advances. First-generation cephalosporins were demonstrated to reliably kill a variety of important bacterial pathogens, and could be used either parenterally (intravenous or intramuscular) or orally. They proved to be remarkably safe. However, the first-generation cephalosporins were not capable of killing several bacteria important in surgical infections, urinary tract infections, and most importantly, septic shock. To fill this void, second- and third-generation cephalosporins were introduced in the West in the late 1970s. Second-generation drugs, such as cefuroxime and cefoxitin, have been used for respiratory infections and intra-abdominal infections, respectively. Third-generation cepha-losporins, such as cefotaxime, ceftazidime, and ceftriaxone, have an even broader spectrum, and somewhat different pharmacological properties. They have been used to treat more severe cases of community-acquired pneumonia, meningitis, and bacterial hospital-acquired infections caused by organisms frequently resistant to other antibiotics (including first generation cephalosporins and other antibiotics).

12.2 The Business of Antibiotics in China, Early 1980s

The first appearance of cephalosporins in China occurred in 1982, when Bristol-Myers Squibb became the first multinational to market a cephalosporin in China, by selling its imported, first-generation cephalosporin, cefradine (brand name: Velosef). In 1985, Bristol-Myers Squibb began producing cefradine at its joint venture plant with Sanwei Pharmaceutical in Shanghai. At that time, the Chinese pharmaceutical industry (like most industries) was fragmented by provinces, as a direct result of infrastructural deficiencies and serious inter-provincial rivalries. In addition, the industry was wholly occupied by state-owned enterprises (SOEs), which had no prior familiarity with, nor need for, modern marketing; no in-house tradition of R&D; no need for modern management skills (as everything prior to this had been politically inspired in a non-market environment); and no abiding interest in the customer (recognizing the historical consequences of several centuries of economies of scarcity). Relying on Michael Porter's "five-forces" model to describe competitive terrain, the Chinese market for antibiotics in the early 1980s was well-portrayed by *Fig. 12.1* (*Porter* 1980).

As *Fig. 12.1* illustrates, there was really no national market, and the essential nature of competition in China at the time was determined province by province. The Chinese logistical infrastructure worked against long-distance transportation, and the absence of commercial media eliminated any chance for the development of advertising and brand-building on a large scale. In addition, and perhaps most importantly, as the provinces developed in a self-reliant fashion, each acquired whatever industrial assets were required for a complete modern economy.

As also indicated in *Fig. 12.1*, none of the five forces effectively possessed any degree of economic influence which might diminish the ability of the competitors in this market to generate profits from their products. While not enjoying the profits of their counterparts in global pharmaceutical markets, Chinese enterprises were, nonetheless, seemingly quite secure in their future, and immune from most of the market forces that were keeping managers in the outside world awake at night. For foreigners, because of molecular exclusivity and the cache of foreign brands, the market was, in the words of one foreign pharmaceutical pioneer in China, "wide open – you could go in and take whatever you wanted."

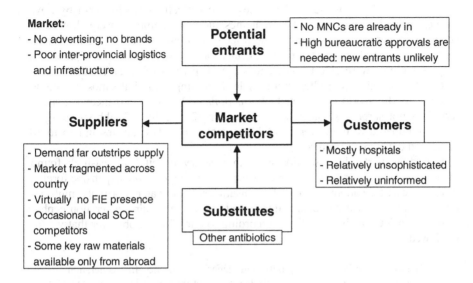

Market:
- No advertising; no brands
- Poor inter-provincial logistics and infrastructure

Potential entrants
- No MNCs are already in
- High bureaucratic approvals are needed: new entrants unlikely

Suppliers
- Demand far outstrips supply
- Market fragmented across country
- Virtually no FIE presence
- Occasional local SOE competitors
- Some key raw materials available only from abroad

Market competitors

Customers
- Mostly hospitals
- Relatively unsophisticated
- Relatively uninformed

Substitutes
Other antibiotics

Fig. 12.1. The cephalosporin market in China in the 1970s and 1980s

In this market, conditions of economic scarcity and local production worked against the building of national brands. Advertising was generally seen as being wasteful of resources that could be better used elsewhere, and technical advances were routinely distributed to all competitors by either the relevant Ministry, or by the State Science and Technology Commission. Price was controlled and standardized nationally for most products, leaving only quality as a way of distinguishing amongst products, and given the imbalance between demand and supply, consumers were quite content to take whatever was available. Prior to the early 1980s, the presence of foreign products in the market was so small as to be of little impact on the Porter forces. As a result, all of the arrows representing the power of Porter's five forces are portrayed as relatively insignificant in *Fig. 12.1*. It might, in fact, be more useful at this period of time to do a five-forces analysis by province, rather than to suggest that a national market existed for all of China. Either way, however, for the reasons given above, antibiotics were an attractive business to be in.

12.3 Market Changes from the Mid-1980s

In the mid-1980s, at nearly the same time as second- and third-generation cephalosporins were emerging in Western markets, the administrative and economic environment around China's health care system was also changing. Hospitals were, and still are, the primary purchasers of drugs in China, purchasing 85 percent of all pharmaceuticals (the remaining 15 percent are purchased by retail pharmacies).

While the fees for most hospital services were typically set below cost by government price bureaus,[1] prices for medicines and new equipment were allowed to be set at 15 percent above cost. These price distortions created strong incentives for hospitals to sell the most expensive drugs and use more medical equipment, as long as patients were able to pay, and since 50 percent of urban residents were medically insured, this resulted in drug sales becoming a hospital's most profitable activity. As the World Bank noted (*World Bank* 1997): With this incentive structure, it is not surprising that drugs account for 52 percent of health spending in China, compared with an average of 14 percent in OECD countries and 15 to 40 percent in developing countries.[2]

Indeed, as market liberalization swept the country, hospitals increasingly found themselves able to generate greater revenues by relying on more expensive foreign drugs, at least for patients with medical insurance. Seeing an opportunity, multinational pharmaceutical companies introduced second- and third-generation cephalosporins into China in the 1990s. According to one Chinese observer that we interviewed:

> Before the 1990s, [China] only had generic, domestic, first-generation cephalosporins. Local doctors only had heard of the newer ones from conferences. Second- and third-generation cephalosporins were too expensive then, and very few doctors requested them – if they had, the pharmacy head would probably have refused. Then...the foreign companies began to lobby the doctors; they then requested the drugs from hospital pharmacies, who ordered from distribution companies, who then bought them from those same foreign companies.

It should not be surprising that antibiotics became the single largest category of purchased items by Chinese hospitals[3], and, in 1996, half the money spent on antibiotics in Shanghai hospitals was for cephalosporins, a percentage that has persisted since then.[4] While some of the most popular antibiotics were older, first-generation cephalosporins, which had been produced as generic drugs by domestic Chinese firms for a number of years, many were the newer (second- and third-generation) products introduced in the 1990s by foreign pharmaceutical companies and their Chinese joint venture partners. Increasingly, domestic companies also began production of these later-generation products as well. These newest cephalosporins were very expensive, often as a result of the need to import costly active ingredients, and, consequently, Chinese regulations limited their use to class II and III hospitals (urban referral hospitals) for serious infections. Nonetheless, they still sold very well around the country: five of the 15 top-selling drugs in

[1] The governmental price system for medical care services is a holdover from earlier days when health care was considered a public good (*Henderson* et al. 1998).

[2] It should be noted that the artificially low prices in other areas would inevitably make the percentage spent on medicines look much higher.

[3] In 2001, "antibiotics accounted for 26% of the total therapeutic area, and in 1999, it was 28-32%." (Confidential industry interview conducted by authors in 2003)

[4] Confidential industry interview conducted by authors in 2001.

all categories, reported in a national sample of high-level hospitals, were the newer cephalosporins. With so many cephalosporin antibiotics introduced in such a short period of time, competition between firms greatly intensified.

By the end of the millennium, the industry had grown in size in both revenues and players, with many more domestic players arising outside of the largest cities and producing generic products at low prices. While foreign firms such as Aventis and Glaxo led the market [in volume and profits, respectively], it was generally thought that the MNCs would not be introducing new cephalosporins into the Chinese market, as there were already a lot of products there, and attention was moving to other classes of antimicrobal agents.[5] The local producers were gaining in sophistication, however, and what the State Development Planning Committee referred to as "global pharmaceutical transfer" could be seen in the origin of the cephalosporins sold in China moving from a roughly 1/3-1/3-1/3 share between local manufacturers, FIEs, and imported products, across all generations of cephalosporins in the mid-1990s, to an increasing share for local producers as the decade ended.

12.4 Characteristics of the Chinese Market for Cephalosporins: Late 1990s

What began as many separate provincial markets in the 1980s, began to take on a slightly less fragmented look in the 1990s (see *Fig. 12.3*). Data collected in 1996 for the sales of two later-generation cephalosporins, *ceftriaxone* and *cefotaxime*, suggest that there was, in fact, in the mid-1990s, the beginning of a national market presence emerging for some enterprises, albeit with significant regional variation. What is particularly significant, however, is that it was not only foreign multinationals who were building a national market presence, but the occasional Chinese producer as well.

However, the basis for competition within this market did not appear to be changing as much as some of the other characteristics of competition. Domestic competitors remained significantly under-priced relative to their foreign rivals. Moreover, the late 1990s saw increased government efforts in the pharmaceutical markets, with the goal of reducing the prices of pharmaceutical products. One consequence of the pressure on pricing, and the already low margins experienced by local firms, is that many domestic competitors now have almost no money to apply to brand-building or innovation. In such instances, price is all that is left, and all that can be done with that is to squeeze the margins even further in what could easily become a spiral into decline.

One of the strategies of the domestic players that appeared to be unfolding as a response to the marketing power of the multinationals was a retreat to the rural markets where the foreigners were not yet present, and where price remained the major factor in the choice of product. The result was a segmentation of the Chi-

---

[5] Confidential industry interview conducted by authors in 2001.

nese market that is depicted in *Fig. 12.2*. Stephen White saw this in his study of the Chinese pharmaceutical industry in general (*White* 1998), and we saw it in cephalosporins, in particular. One Chinese enterprise told us that by moving to the rural areas, it grew its market by 40 percent!

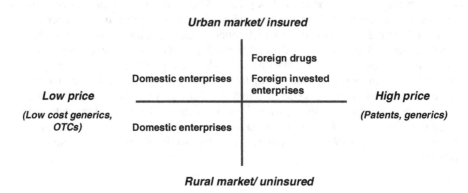

Fig. 12.2. Recasting the Chinese market: segmentation

William Keller, General Manager of Keller Pharma Consultancy (Shanghai), and the former Managing Director of Roche China, who is one of the most astute observers in the foreign expatriate community in China, describes this market segmentation phenomenon by quoting Mao Zedong: "Occupy the countryside. Surround the cities. Win the War." According to Keller:

"This is the strategy that Chinese companies are using. We, Western firms, are focused on tier 1 cities. They go to tier 2 and 3 cities, learn from us, and then attack. As a result, even though Western firms have made rapid inroads, the tide is changing. You must not underestimate Chinese companies.

We are racing to learn about China and they are learning about market economies – a race in parallel – and he who reaches their learning objective first wins all."[6]

Price could therefore be seen as an alternative to scientific innovation for local players in the Chinese market. Despite market-share inroads by Chinese domestic firms in later-generation cephalosporins, there was a strong feeling among most observers, foreign and Chinese alike, that very little scientific innovation was being undertaken within Chinese pharmaceutical firms. While R&D activities might be scarce, evidence of non-technical innovation, such as imaginative approaches in marketing or distribution, could be seen in a number of domestic firms. In the Shanghai market, New Asiatic Pharmaceutical, for example, employed competitive strategies that it had learned, in part, from companies like Roche and Squibb to develop its own brand. They also created sales teams involving physician-

---

[6] William Keller, comments made during IMD EMBA discovery visit to Shanghai, October 2003.

representatives to build relationships with key decision-makers, something that the foreign-invested enterprises [FIEs] had long done as part of their marketing efforts. It is also instructive that in the Chinese markets for *ceftriaxone* and *cefotaxime*, while foreign firms predominated in the markets that they were in, Chinese domestic competition did not collapse in the face of the foreigners, despite the low margins and competitive pressures that they faced.

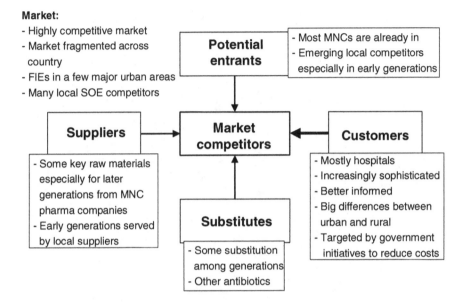

Market:
- Highly competitive market
- Market fragmented across country
- FIEs in a few major urban areas
- Many local SOE competitors

Potential entrants
- Most MNCs are already in
- Emerging local competitors especially in early generations

Suppliers
- Some key raw materials especially for later generations from MNC pharma companies
- Early generations served by local suppliers

Market competitors

Customers
- Mostly hospitals
- Increasingly sophisticated
- Better informed
- Big differences between urban and rural
- Targeted by government initiatives to reduce costs

Substitutes
- Some substitution among generations
- Other antibiotics

Fig. 12.3. The cephalosporin market in China in the late 1990s

In fact, what appears to have happened in cephalosporins is not so different from the competitive patterns that were played out in China in the early part of the twentieth century. During that period, in quite a few industries, well-endowed international players, with widely respected brand names, and world-class quality and design, entered the Chinese market, only to be surprised (*Cochran* 2000). They bought local brands, they built national distribution systems, they established FIEs to serve as their local presence, and they expected the market to fall into their laps. Of course, many local competitors succumbed almost at once to these invasions, but not all. In the face of strong foreign competition, a variety of interesting domestic responses arose. In many industries, there was a consolidation of domestic producers in order to strengthen the survivors. Often, the domestic producers would retreat from the major urban markets, to buy time for rebuilding by serving the rural markets that were much more difficult for the foreigners to access. Frequently, they would appeal to patriotic or ethnic impulses to win customers back from the foreign interlopers. That is what appeared to be happening in China's cephalosporin market in the late 1990s.

As China's rural markets grow in sophistication, and as domestic producers grow in competitive strength, we can expect to see pitched competitive battles fought around the fringes of the major urban areas, as a prelude to the eventual battle for the cities. This, of course, is what Keller was referring to when he quoted Mao Zedong.

Whether or not this presages eventual Chinese competition on the global cephalosporin stage, however, remains to be seen.

Part III: The Experience of Successful European Companies in China

13 Swimming Ahead of the Shoal – The History of BASF in Greater China

Andreas Kreimeyer: BASF Aktiengesellschaft, Ludwigshafen, Germany

There is no truer business principle than striking the right market fast, and staying with it. This says a great deal about a company's foresight and its ability to weather change. BASF's 120-year history in China has attested to this principle.

China's remarkable economic and social progress since the reforms of 1978 has brought about an almost unparalleled, successful transformation of the country and its position in the world. Two events in particular have galvanized national sentiment and signaled China's arrival on the world stage. In 2001, the International Olympic Committee awarded the 2008 Games to Beijing. And in 2003, astronaut Yang Liwei spent 23 hours in orbit, making China only the third country to achieve manned space flight. The IOC decision demonstrates the confidence that the international community feels towards China. China's emergence as a space power, meanwhile, is evidence of the country's own confidence, as well as a growing optimism about the future – a future that is equally promising to international business players including BASF.

When BASF first began trading in China in 1885, the country, ruled by the crumbling Manchu dynasty, was struggling with the challenges presented by outside forces increasingly determined to open new markets and compel China to abandon its isolationist policies. Faced with the same pressures that had forced a reluctant Japan to open the door to international trade and institute a modernization program, the Qing dynasty tried instead to maintain its position of minimal contact with the outside world. It was not until the revolution of 1911 that the first hesitant steps were taken to reverse this course.

Like other European powers in the late 19th century, Germany was determined to explore the potential of what was believed to be a vast, largely untapped market in China. Although Portugal had the longest-established permanent presence in the region, having established a colony at Macau in the 16th century, it was the British takeover in 1841 of Hong Kong, with its fine natural harbor, which accelerated the process of foreign expansionism in China. Treaty ports were set up, along with trade zones and foreign concessions in a number of cities.

13.1 The Founding of BASF

BASF was founded by Friedrich Engelhorn, the owner of a company in Mannheim which produced gas for illumination. Engelhorn was aware of the recent discovery that coal-tar – a by-product of gas manufacture and therefore readily available to him – could be used to produce dyestuffs. He began making aniline and fuchsin in 1861, and in 1865 set up a specialized company for large-scale manufacture: Badische Anilin- und Soda-Fabrik (BASF). Unable to secure a site in Mannheim, he located the new company across the Rhine in Ludwigshafen.

BASF quickly became a leading producer of coal-tar dyes. Discovered in the 1850s and displayed to public acclaim at the World Fair in London in 1862, it soon became apparent that such synthetic dyes could be produced more cheaply and effectively than their vegetable-based counterparts. They were also more durable and reliable. It was clear that there was exceptional potential for profit, with virtually unlimited demand from the rapidly expanding textile industry. Manufacturers scrambled to obtain the licenses and technology for the new process.

13.2 A Partnership Sealed by Colors

The success story of BASF in China was destined to be sealed by colors, too. Expansion of its sales organization prompted the company's first activities in China in the mid-1880s, when it sold and marketed the magenta dye fuchsin to the flourishing cotton cloth market. By 1900, China was the world's largest importer of cotton yarn, and the second largest importer of cloth. The Shanghai office of Ehlers & Co. was appointed agent for northern China, and Theodor Meyer was sent to Ehlers as BASF's representative. Stoltenfoot & Hagen in Hong Kong was engaged as agent for southern China until 1897, when Jebsen & Co. took over.

In 1897, after more than a decade of intensive and costly research, BASF succeeded in producing a synthetic indigo. Its development had taken 17 difficult years and consumed around 18 million gold marks – more than the entire stock capital of the company at the time. The new product was something of a 'Holy Grail'. As a substitute for one of the most widely used natural dyestuffs, it was expected to be immensely profitable. In the early years of the 20th century, China, where 'almost every [peasant family] owned at least one blue jacket dyed with indigo', was seen as potentially the most lucrative market. At the same time, as with other chemical dyes, convincing people to switch from the traditional product was not always easy, especially in areas with limited communication with the outside world. It was not uncommon for families in the countryside to dye their own clothes, so that while synthetic indigo quickly caught on in the cities of the eastern coastal regions and overtook the natural product, it gained acceptance much more slowly in the hinterland. As transport infrastructure and communications gradually improved, it was hoped that increased sales would follow.

Between 1890 and 1904, there was a significant improvement in sales of BASF products to Asia, which until then had accounted for only 4 percent of turnover. The surge was mainly owing to the fact that large quantities of synthetic indigo were being purchased as a substitute for the inferior natural product. Performance in China mirrored that of Japan, each country accounting for around 7 percent of BASF's total worldwide turnover by 1904. This severalfold increase came at a time when in Europe, sales had slumped, notably in Britain, which was one of the company's most important markets after Germany itself.

13.3 The Two World Wars

In 1914, German production accounted for 85 percent of the world dyestuffs market. Despite a British blockade of German ports, for the first year after the outbreak of the First World War BASF did manage to continue sales to China, using stock that was already outside Germany at the onset of hostilities and shipping from ports in neutral countries such as Portugal.

At the end of the war, under the terms of the Versailles treaty, Germany was committed to paying punitive reparations, which would partially have to be met through industrial output. The setbacks that had been experienced by BASF were partly alleviated by renewed demand from Asia for German dyestuffs, still seen as superior in quality to those produced by rival countries. The year 1920 saw a dramatic increase over prewar sales in China: 67,000,000 marks in the first quarter, against just 17,000,000 marks for the whole of 1913. However, the market was quickly oversupplied, and a slump followed. With the decline in revenue from indigo during this period, BASF sought to diversify its range of products in China. In the early 1920s, indanthrene dyes were introduced, and towards the end of the decade, BASF began selling nitrate fertilizers. The first big order was for 5,000 bags of ammonium sulfate, destined for the tea plantations near Fuzhou. Fertilizers were also sold in Taiwan, still a Japanese colony at the time, through the Mitsubishi Trading Corporation.

The 1930s were ominous years in both Europe and Asia. The rise of Nazism ended hopes for democratic change in Germany, and a second confrontation in Europe moved from a possibility to a certainty. Meanwhile, Japan's expansionism, particularly in China, eventually pulled Asia and the whole Pacific region into conflict, and the war became global.

13.4 The Postwar Era

Although China emerged from the Second World War as a victor, having defeated the Japanese occupation, its position was hardly less precarious than Germany's. Loss of life had been enormous, infrastructure was severely damaged and, worst of all, enmity between the Kuomintang and the Communists, set aside during the fight against the Japanese, boiled over once again into military conflict. The ensuing civil war came to an end in 1949 with the founding of the People's Republic of China. The Kuomintang retreated to Taiwan. Whether as a result of the lack of diplomatic recognition for the PRC or whether it was simply Communist policy and the consequent shift in international allegiances, direct trade between China and the capitalist powers effectively ceased in the following decades. The manufacturing and investment focus switched to Taiwan, while Hong Kong built up its position as China's gateway to the outside world. Shanghai, once Asia's most international city, was still China's main port and greatest metropolis, but compared to Hong Kong's new-found vibrancy, it was entering an era of stagnation and decline that would last several decades.

The difficult postwar years of the early 1950s did not drive BASF away from China, but rather provided some recuperation time for the company while it collected insight into trends in fashion and information on customers' wishes. While still accumulating its local wisdom, BASF had appointed Hong Kong-based Jebsen & Co. as its sole agent for China trade. Jebsen managed to maintain its offices in Shanghai and Tianjin during the early years of the People's Republic. The firm continued to represent BASF from Hong Kong, until BASF began its own direct engagement with China in 1982.

After 1945, BASF exported fertilizers to Taiwan through Jebsen & Co. In the early 1950s, BASF opened an office with United Exporters in Taipei and then with Delta Inc. towards the end of the decade. BASF's first direct investment in Taiwan came in 1969 when it established the Cheng Kuang Chemical Industrial Co. Ltd and Teh Hsin Dyes and Chemicals Co. Ltd. Cheng Kuang was the manufacturing arm, while Teh Hsin was a trading company. Cheng Kuang became BASF Taiwan in 1984 and merged with Teh Hsin four years later, after Taiwan's government restrictions on foreign companies were eased.

By the 1980s, China was becoming more open as the result of the ambitious reform program instituted by Deng Xiaoping following the death of Mao Zedong. The door to capitalist investment had been reopened, and BASF was quick to seize the opportunity.

Well aware of the fact that to get into a market you have to be local, in 1982 we took our first step in strengthening our presence in China by establishing a subsidiary in Hong Kong, BASF China Ltd, with responsibility for sales, marketing and distribution of our products in Hong Kong and on the mainland. A representative office was opened in Beijing in 1986, followed by offices in Shanghai, Nanjing, Guangzhou, Qingdao and Chengdu.

Mr. Klaus-Dieter Preuss, the first Managing Director of BASF China, who was personally involved in the research and setting up of the company, confirmed the timeliness of the decision to be one of the first chemical companies to set up offices in China. "When I left BASF China in 1990, sales turnover was five times what it had been when the company started in 1982, and growth was still surging."

In 1994, China relaxed controls on foreign participation in the chemical sector in the hope of attracting US$ 10 billion in investment and technology by the end of the decade. While players in the industry were skeptical about the profit potential of such investments, BASF had already established three production facilities in China, making it the second-largest German investor in the country. The first of BASF's joint ventures was set up in 1988 in conjunction with the Shanghai Gao Qiao Petrochemical Corporation - Shanghai Gaoqiao BASF Dispersions Co. Ltd (SGBD) - to produce styrene-butadiene dispersions for coating paper and carpets to supply rapidly expanding customer markets. Since then, we have successfully established other joint ventures, expanding the local production base of BASF in Greater China (see *Table 13.1*).

Table 13.1. Expansion of local production in Greater China

Year of establishment/ Location	Name of production facility	Product	Partner	Remark
1988/ Shanghai	Shanghai Gaoqiao BASF Dispersions Co. Ltd (SGBD)	Styrene-butadiene dispersions for coating paper and carpets	Shanghai Gao Qiao Petrochemical Corporation	
1988/ Hsinchu, Taiwan	BASF Polyurethanes (Taiwan) Co. Ltd (BAPT) (previously BASF Headway Polyurethanes (Taiwan) Co. Ltd)	Polyurethane system	Headway Group	Became BASF's wholly-owned subsidiary in 2002
1994/ Shanghai	BASF Auxiliary Chemicals Co. Ltd (Shanghai BASF Colorants and Auxiliaries Co. until 2000, BASF Colorants and Chemicals Co. Ltd until 2004)	Organic pigments and metal complex dyes, leather and textile auxiliaries, acrylic dispersions, acrylic copolymers and blending of fire-resistant hydraulic fluids	Shanghai Dyestuff Co.	Became BASF's wholly-owned subsidiary in 2000
1994/ Nanjing	Yangzi-BASF Styrenics Co. Ltd (YBS)	Ethylbenzene, styrene, polystyrene and Styropor®	Yangzi Petrochemical Co.	
1995/ Shanghai	BASF Shanghai Coatings Co. Ltd (BSC)	Coatings, such as cathodic electro-coats, primer surfacers, topcoats, basecoats, clearcoats, auxiliaries and plastic coats, mainly for the automotive industry	Shanghai Coatings Co. Ltd	
1995/ Jilin	BASF JCIC Neopentylglycol Co. Ltd (BJNC)	Neopentylglycol (NPG)	Jilin Chemical Industrial Co. Ltd	

Table 13.1. (continued)

Year of establishment/ Location	Name of production facility	Product	Partner	Remark
1995/ Shenyang	BASF Vitamins Co. Ltd (BVC)	Vitamins, vitamin blends and non-vitamin feed additives	North East General Pharmaceutical Factory	BASF's shareholding increased from 70 percent to 98 percent in 2000
1997/ Guangzhou	BASF Polyurethanes (China) Co. Ltd (BAPC) (previously BASF Headway Polyurethanes (China) Co. Ltd)	Polyurethane systems	Headway Group	Became BASF's wholly-owned subsidiary in 2002
2000/ Nanjing	BASF-YPC Co. Ltd. (BYC)	High-quality chemicals and polymers	China Petroleum & Chemical Corp. (Sinopec)	Scheduled for commercial operation in mid-2005
2002/ Shanghai	BASF Chemicals Co. Ltd. (BACH)	Polytetrahydrofuran (PolyTHF®) and tetrahydrofuran (THF)	N/A	BASF's wholly-owned subsidiary; scheduled for mechanical completion in the first quarter of 2005
2002/ Hsinchu, Taiwan	BASF Tai Ching Crop Protection Chemicals Corporation (BTCC)	High-quality crop protection formulation products	Taiwan Sugar Corporation	
2003/ Shanghai	Shanghai BASF Polyurethane Co. Ltd. (SBPC) & Shanghai Lianheng Isocyanate Co. Ltd. (SLIC)	Crude MDI (diphenylmethane diisocyanate) and TDI (toluene diisocyanate)	Huntsman, Shanghai Hua Yi (Group) Company, Shanghai Chlor-Alkali Chemical Co. Ltd. and Sinopec Shanghai Gao Qiao Petrochemical Corporation	Scheduled for commercial operation in 2006

By 1995, around 700 staff were employed in BASF's China operations. We had sold products worth DM 660 million (US$ 440 million) the previous year, of which almost 10 percent were manufactured in China itself. Since the early 1980s, chemical consumption in the PRC had grown at the rate of 8 percent a year, and the market, worth US$ 48 billion, was projected to more than triple by 2010. China's importance as a market was instrumental in the decision, in 1995, to base BASF's East Asia Regional Headquarters in Hong Kong. It had responsibility for Greater China (the PRC, Taiwan and Hong Kong), South Korea and Japan. For the first time, a member of the Board of Executive Directors, Juergen Hambrecht, was based in Asia, underlining the strong role BASF intended to take there. The primary function of our new headquarters was to oversee operations in each of the countries and develop new projects for the BASF Group. The following year, a holding company, BASF (China) Co. Ltd, was formed in Beijing with responsibility for integration of BASF's mainland China operations and provision of corporate services to joint ventures, encompassing human resources, accounting, financing, logistics, reporting, IT, marketing and sales.

Upon becoming President of BASF East Asia, Juergen Hambrecht quoted the three-word mantra central to BASF's "Vision 2010", "change, focus and speed", as the success factors for BASF's approach to the market. "We all have to change. Investing in East Asia, and especially China, needs high flexibility and steady adaptability to respond to the big challenges of working across cultures in order to meet the specific needs of the market. And we have to change from an opportunistic market approach to systematic and sustainable market penetration based on domestic manufacturing facilities."

In recognizing that change was the rule rather than the exception, BASF set out in 1996 to define clear targets for the whole of Asia Pacific in terms of where we wanted to be at the end of the first decade of the 21st century:

- Participate in high growth markets;
- Contribute 20 percent to BASF Group sales and earnings in the chemical business;
- Strengthen position as one of the top five suppliers in strategically relevant markets;
- Build a local manufacturing base that will secure 70 percent of regional turnover;
- Form the best team in the Asia Pacific chemical industry.

At the time of setting these targets, there was much work to be done if they were to be reached. Local production, for example, would have to double. As part of this ongoing process, our company aims, where possible, to delegate responsibility for activities and operations from headquarters to the region. Establishing the regional headquarters in Hong Kong and the holding company in the PRC is an important contribution to our strategy. This decentralization allows for a quicker and more efficient approach to meet the requirements of customers.

1997 was something of a watershed year in the region, marked most prominently by the Asian economic crisis, which was precipitated by currency devaluations in Thailand and Korea. By 1998 the shockwaves had spread throughout Asia. One country, though, seemed miraculously unscathed: China, whose economy – according to official statistics – continued to grow at an impressive rate, while Taiwan and Hong Kong both languished. Despite the slump, we pushed ahead with ambitious expansion plans for the region, including a vast petrochemical project in Nanjing. The crisis in Asia Pacific was taken by us as a chance to enforce necessary structural reforms and refocus on sound economic development. At BASF, we did not revise our ambitious targets for Asia Pacific.

BASF had already decided that Asia was the key to future growth. The Asian economy was expanding at four times the rate of the Western markets. It was responsible for 12 percent of BASF Group sales, and our company had increased our business by 23 percent in the previous year. There had been a record profit for the BASF Group worldwide and a 10-percent increase in Asian sales. BASF intended to invest up to US\$ 3 billion in Asia over the next five years, which constituted 25 percent of BASF Group's worldwide total. China was the main focus in the region. Two joint ventures had just come on stream: Shanghai BASF Colorants and Auxiliaries Co. had begun production of textile dyestuffs and pigments in 1996, and BASF Shanghai Coatings Co. had started manufacturing and marketing coatings and paints in early 1997. Yangzi-BASF Styrenics Co. Ltd became operational by fall 1997. In addition, there was the Nanjing project on the drawing board.

Together with China Petroleum & Chemical Corp. (Sinopec) in a 50:50 partnership, we wanted to build our first Verbund project in East Asia – an integrated petrochemical site (IPS) on 220 hectares of land. The core of the project was an ethylene cracker with a capacity of 600,000 tons per year. Nine new plants downstream, part of the same project, would be supplied by the cracker, producing 1.7 million metric tons of chemical products for local consumption, including ethylene, aromatics, polyethylenes, ethylene oxide and ethylene glycols, acrylic acid, acrylates, oxo alcohols, formic acid, propionic acid, methylamine and dimethyl formamide.

The integration, or Verbund, concept has a history stretching back to the early days of production at the Ludwigshafen site in the late 19th century, and today it is our core competence. By-products or waste materials from one manufacturing process can in turn be used in another, increasing efficiency and cutting costs. While the idea was acknowledged and used at BASF plants a century ago, it was honed into a defining corporate strategy in the early 1960s. What is envisaged is a highly sophisticated and flexible chain of production, where plants are effectively interlocked, and the optimum use of resources, energy and manpower can be achieved. For example, a group of sites built in close proximity reduces transport costs; the heat given off during manufacturing at one plant is used to provide energy to another. A further beneficial effect besides cost and efficiency is the potential for reducing energy requirements and waste generation. Over the years, the Verbund concept has evolved to cover a wide range of activities which can benefit from better coordination and integration. Examples include R&D, purchasing, logistics, and BASF's interaction with customers and the community at large.

The Verbund strategy is the core strength of BASF's activities worldwide. The first Asian Verbund site was built at Kuantan in Malaysia, to serve the Southeast Asian market. The Nanjing or IPS project, boasting the largest construction endeavor in BASF's history, with around 14,000 workers at the peak of its construction, required an investment of US$ 2.9 billion. Strategically located at one of China's chemical bases in the Yangtze River Delta and well supported by an extensive rail network and other logistics infrastructure, our project will mainly supply local customers in the prosperous eastern coastal cities.

After gaining approval from the State Council – the first time that a cracker project with foreign participation had done so – the joint venture contract between BASF and Sinopec was signed in Berlin in June 2000 at a ceremony attended by German Chancellor Gerhard Schroeder and Chinese Prime Minister Zhu Rongji. The groundbreaking ceremony took place in late September 2001, at what was a highly uncertain time globally, with the United States still in shock from the attacks on New York and Washington. To some, it must have seemed a highly inauspicious time to be embarking on such a course. However, while security was certainly uppermost in people's minds in all countries and there was a realization of the vulnerability of major installations and industrial sites, realistically, there was no likelihood of a severe long-term impact on economic or industrial activity. We vowed to remain committed to our investment decisions and would continue to explore new projects in and for the Asia Pacific region and, in the future, would not miss any opportunity to grow even stronger.

The Herculean scale of the Nanjing project requires courage, self-confidence, enthusiasm and unlimited commitment on the part of all those involved: employees, partners, contractors and suppliers. However, we strongly believe the growth forecasts for the chemicals market in the region – and for China's gross domestic product, which is predicted to increase at the rate of 7 percent a year.

13.5 Towards the Future

Certainly this approach can now be seen to be paying off. BASF made sales in Greater China in 2003 of close to EUR 1.6 billion, almost 20 percent of which was from local production. Since 1996 we have grown our business in Greater China by a compound annual growth rate of more than 20 percent (see *Fig. 13.1*).

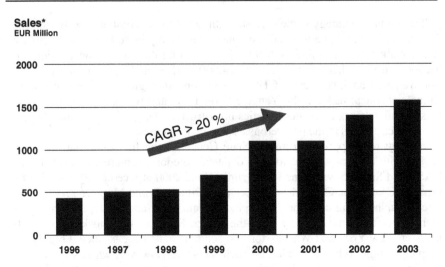

Sales*
EUR Million

* Sales Greater China to third parties by location of customers in 2003

Fig. 13.1. Strong business growth since 1996

The greater portion of sales was in plastics, which accounted for 54 percent, followed by performance products (20 percent), chemicals (18 percent) and agricultural products and nutrition (7 percent). Our policy is to localize production as far as regulations allow, to which end we now have ten wholly-owned enterprises in Greater China, and nine joint ventures (see *Fig. 13.2*).

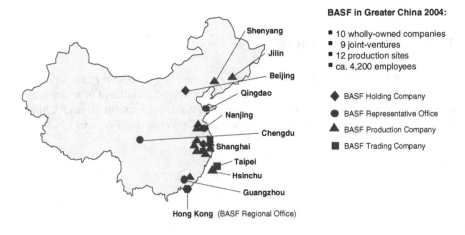

BASF in Greater China 2004:

- 10 wholly-owned companies
- 9 joint-ventures
- 12 production sites
- ca. 4,200 employees

◆ BASF Holding Company
● BASF Representative Office
▲ BASF Production Company
■ BASF Trading Company

Fig. 13.2. Established business network in Greater China

Our company's workforce in Greater China has grown to more than 4,000, and we predict that this figure will surpass 5,000 within a few years. The Nanjing petrochemical complex is due to begin operation in mid-2005. The world's largest integrated production facilities for polytetrahydrofuran (PolyTHF®) and tetrahydrofuran (THF) at Caojing, Shanghai will be operational in 2005. Here, butane will be converted directly to THF and subsequently to PolyTHF®, eliminating the previously necessary intermediate step via 1,4-butanediol (BDO). It is the first location worldwide where the process, which was developed by BASF, is being utilized on a commercial scale. 60,000 metric tons of PolyTHF® and 80,000 tons of THF per year will be produced for the spandex fibers market, mainly for use in sportswear and garment manufacture. Also at Caojing, we are collaborating with Huntsman Polyurethanes and our Chinese partners to build a plant which will manufacture 240,000 metric tons of crude MDI and associated precursors, an MDI finishing plant, and a TDI complex producing 160,000 tons per year and associated precursors. Scheduled to begin operation in 2006, these plants require an investment of US$ 1 billion. BASF's commitment to Greater China has been demonstrated by these world-scale projects and the massive investments that underpin them.

Before retiring from the chairmanship of BASF, Juergen Strube visited Beijing in October 2002 and told the press there: "We are investing in more than just steel and pipes in China. Here, as in other parts of the world, our operations are guided by the principles of sustainable development."

Indeed, sustainable development takes the form of values and principles within our company, enhanced through a management system that defines each and everyone's role in the company and speaks for itself through action. The three pillars of economic development, environmental protection and social responsibility that underpin the sustainability principle have been exemplified through our activities in China. To support Chinese universities and research institutes in achieving scientific excellence, the BASF Sino-German Research and Development Fund was established in 1997. Since then, it has supported over 40 research projects and benefited close to 1,100 students from ten universities. It now forms an important part of BASF's research and development efforts in China, which are seen as a key to fully realizing the goal of integration. In January 2004, BASF became a founding member of the China Business Council for Sustainable Development. We are proud to be part of this initiative, to share the same vision and concern as the other founding members including Sinopec, which is now not only our investment partner but also a partner on the journey to promoting sustainable development in China.

Another crucial aspect of future growth is attention to environmental matters. Clearly for China, this is of vital importance, for several reasons. Pollution problems and environmental degradation are already severe. Despite some encouraging signs of a desire on the part of the authorities to deal with the problems of carbon dioxide emissions, poor air and water quality, deforestation and soil erosion, there are obstacles to overcome before the paradox of increasing economic development while protecting the environment can be resolved. BASF has played a part here, too. In June 2004, we signed an agreement with the Chinese Research Academy of Environmental Sciences to jointly set up an engine testing laboratory in Beijing.

The laboratory will be responsible for setting and supervising product standards related to fuel quality, assessing the quality of Chinese gasoline, improving the quality of fuels for motor vehicles and thus reducing emissions. BASF also offers a whole line-up of products that benefit the environment in China. These include Keropur®, a gasoline additive that helps reduce emissions; Ecoflex®, a biodegradable polymer that addresses the problem of plastic waste; Cyclanon® ECO, a postclearing agent for use in the textile industry that helps save water.

Harnessing the changing business environment for sustainable, profitable growth, we announced our renewed strategy 2015 in late 2003, which set the path to the future for our subsidiaries and affiliates. Four strategic guidelines shape who we are and what we stand for:

- Earn a premium on cost of capital;
- Help customers to be more successful;
- Form the best team in industry;
- Ensure sustainable development.

As an important growth engine of BASF in Asia Pacific, Greater China represents a key contributor in helping achieve the company's goal to remain the world's leading chemical company.

Our production in Greater China is both diverse and comprehensive: polymer dispersions, pigments, styrene, polystyrene, polyurethanes, engineering plastics, coatings, finishing products for the textile and leather industries, intermediates, vitamins and crop protection products. With increasing personal wealth in the PRC, the gap between European and Chinese patterns of consumption is closing fast. Few people can now doubt that the next century in China will change the world. Economically, socially and politically, the country looks set to evolve more swiftly than any other in recent history. While it is impossible to predict the exact course of events, no one wants to miss out on playing a part. BASF has taken up the challenge, and set its course. Our projects in Asia are making good progress, and we are determined to sustain our leading position in the market. In doing so, we will not confine ourselves to the expectations of the market, but outperform them. As a popular Asian saying goes: "It's better to swim ahead of the shoal than to drift with the current."

14 Establishing a Competitive Production Network in Asia

Otto Kumberger: BASF Aktiengesellschaft, Ludwigshafen, Germany

14.1 Asia – Shaping the Future of the Chemical Industry

In the 1980s, Asia was still considered by many as a backwater in the global chemical industry. A decade later, the potential of the Asian markets was discovered and chemical companies started to invest in Asia. Today we are all convinced that Asia will shape the future of the chemical industry. Already, the demand for chemicals in the Asia-Pacific exceeds that of Europe and approaches the size of demand in NAFTA. In 2002 Asia-Pacific accounted for 31 percent of global chemicals demand (*Fig. 14.1*). We expect that the Asian chemical industry will grow at an annual rate of almost 4 percent up until 2015, continuing to outperform other regions and increasing its share of global chemicals demand to 34 percent in 2015. As a consequence, the global demand for chemical products will in the future be even more strongly concentrated in Asia. The main driving force for this development is the rapid growth of a strong industrial base, including the migration of manufacturing industries from Europe and NAFTA to Asia.

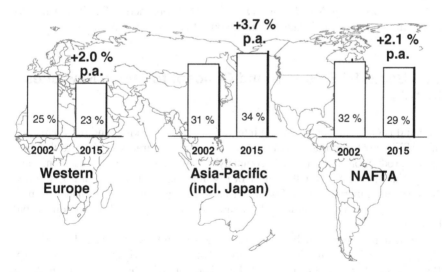

Fig. 14.1. Structure of the global chemical market (excluding pharmaceuticals) 2002 and 2015

To serve these markets *locally*, BASF has built a *tailor-made network* of 44 production sites generating over 50 percent of BASF's sales in the region today, with a clear target to increase that figure to 70 percent by 2015 at the latest. Competitive chemical production on the basis of world-class technology and economies of scale is a key success factor for a chemical company. In this chapter we would like to explain BASF's production network and the outstanding role of BASF's Verbund concept.

14.2 BASF's Production Philosophy

BASF's production network in Asia consists - in line with the global production philosophy - of Verbund sites, chemical sites and sites close to customers.

It centers on two flagships, the *Verbund sites* in Nanjing, China and Kuantan, Malaysia. In BASF's terminology, a Verbund site consists of several interconnected value chains with economies of scale. As well as Verbund sites, BASF operates six so-called *chemical sites* (Yeosu and Ulsan, Korea; Caojing, China; Yokkaichi, Japan; Singapore; and Mangalore, India). Chemical sites are integrated sites with world-scale plants. Compared to a Verbund site, however, the scope is in most cases focused on one value chain only.

Lastly, BASF operates several sites *close to our customers*, where such proximity is critical for success because logistics costs and timing outweigh the advantage of being part of an integrated site. Typical examples are for products such as dispersions, paper chemicals, textile chemicals, polyurethane systems and vitamin premixes.

Overall, BASF relies on a fine-tuned network of production facilities, which is constantly optimized and extended in response to market requirements.

14.3 Using BASF's Verbund Concept to Shape New Sites in Asia

Verbund is German word meaning integration and networking, and represents one of BASF's most important strengths. BASF's Verbund traces its roots to the integrated production system at the company's original site in Ludwigshafen, Germany. Starting from a small number of raw materials (such as natural gas, crude oil, ores and minerals), about 200 major basic products and intermediates are produced. These chemical building blocks are the basis of some 8,000 products sold by BASF.

The product range can be compared to a tree and its branches, the "chemis-tree" (*Fig. 14.2*). At our European Verbund sites in Ludwigshafen and Antwerp, Belgium, crackers are the 'roots' of the "chemis-tree." They act as raw material suppliers for a branched production network. The crackers convert the petroleum fraction naphtha into basic petrochemical products. Starting from these base products, higher value products are manufactured in interconnected, branching value chains. Managing such integrated value chains is BASF's core competence.

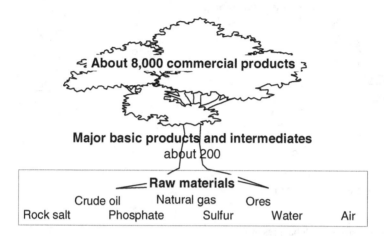

Fig. 14.2. The "Chemis-tree"

By integrating production in a network of plants, the products and by-products of one plant can serve as the raw materials in other plants (production Verbund). Integrated production helps to conserve energy and resources (energy Verbund). BASF has subsequently expanded the Verbund strategy across the technical area. It is reflected, for example, in BASF's coordinated approach to purchasing and research, where our activities are integrated in central competence centers (know-how Verbund). And the Verbund does not stop with the company. It also includes BASF's business partners and customers.

Verbund is not only a theoretical concept, but also an important asset for generating solid financial benefits. Calculations show that the Verbund saves BASF about EUR 500 million each year at the company's Ludwigshafen site alone (EUR 300 million for logistics, EUR 150 million for energy and EUR 50 million for infrastructure). As well as reducing production costs, BASF's Verbund also protects the environment by helping to reduce waste and cut emissions while keeping the consumption of resources and transportation to a minimum. The Verbund concept is used throughout BASF's global operations. With regard to production, BASF today has six Verbund sites worldwide, which form the backbone of our manufacturing activities (*Fig. 14.3*): Ludwigshafen and Antwerp in Europe, Freeport and Geismar in the United States, Kuantan on the east coast of Malaysia and Nanjing at the heart of China (start-up in 2005).

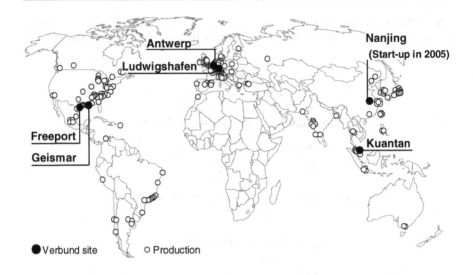

Fig. 14.3. Verbund - global presence

14.4 BASF's Investments in Asia-Pacific

BASF is one of the largest foreign investors in Asia. Key targets of BASF in Asia are:

- strengthen position as one of the top five suppliers in strategically relevant markets;
- to contribute 20 percent to BASF's group sales and earnings in the chemical business;
- to build a local manufacturing base that will secure 70 percent of regional turnover;
- to form the best team in the Asia-Pacific chemical industry.

The basis for building up a local manufacturing base is BASF's ongoing investment program. By 2005, BASF's total investment in Asia-Pacific will be worth EUR 5.6 billion. Key investments are (*Fig. 14.4*):

- our new world-scale Verbund sites in Nanjing, China, and Kuantan, Malaysia;
- a styrene monomer and propylene oxide (SMPO) plant on Jurong Island, Singapore (on stream since 2002);
- a PolyTHF®/THF project in Caojing, China;
- a MDI/TDI complex at the same location;
- major investments at Yeosu and Ulsan, South Korea.

Fig. 14.4. BASF in Asia-Pacific: key investments

14.4.1 Kuantan – BASF's First Verbund Site in Asia

Kuantan is BASF's first Verbund site in Asia. It has been developed together with our partner Petronas, the Malaysian oil and gas company. The joint venture company BASF PETRONAS Chemicals Sdn. Bhd. was founded in 1997. BASF and Petronas hold 60 percent and 40 percent of the company, respectively. The first production complex (acrylic monomers) became operational in mid-2000 (*Fig. 14.5*). In April 2001 the second complex (oxo/syngas/phthalic anhydride/plasticizers) followed, and at the beginning of 2004 the third complex (butanediol) successfully started operation.

As a further extension of our Verbund in Kuantan we plan to construct a plant for polybutylene terephthalate (PBT) together with Toray Industries, Japan.

The main raw material used at the Verbund site in Kuantan is propylene as the starting material for the C3 value chain. It is provided by a propane dehydrogenation unit operated by our partner Petronas. Syngas and hydrogen are produced from natural gas, which is also supplied by Petronas.

Propylene is used as the starting material for the acrylic acid value chain (acrylic acid and the downstream products glacial AA, butyl acrylate and 2-ethylhexyl acrylate) as well as for the production of C4 oxo alcohols (main products: n-butanol and 2-ethylhexanol). The C4 oxo alcohols are also sold in the market but are mainly used as starting materials for the production of acrylates (butyl acrylate, 2-ethylhexyl acrylate) and plasticizers. The second component needed to manufacture plasticizers, phthalic anhydride, is also produced on site on the basis of o-xylene. In the third complex, n-butane is oxidized to form maleic anhydride, which is subsequently hydrogenated to yield butanediol. A downstream extension of this value chain will be the planned PBT plant. Syngas and hydrogen produced in the syngas plant are used in the oxo and butanediol complexes.

Main Verbund aspects are the product integration at the site, where an interlinked network of value chains secures critical mass for the main components (e.g. oxo alcohol and syngas plants). A particular characteristic of our Kuantan Verbund is the close integration with a partner: propylene, natural gas and utilities (central utility facility) are provided by our partner Petronas.

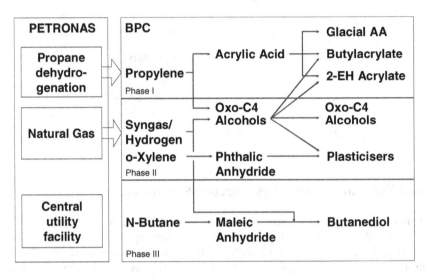

Fig. 14.5. Product Verbund at Kuantan (simplified)

14.4.2 Nanjing – A Cornerstone in BASF's Development in Asia

BASF's second Verbund site in Asia is currently built in Nanjing, at the heart of the Chinese market which is currently the fastest-growing major chemical market in the world. This growth has included high investment in chemical plants, many of them in newly developed chemical parks. Today China has probably the highest concentration of chemical parks worldwide. Just taking into account major sites, an impressive number of about 20 chemical parks are scattered around China offering a total available area of more than 30,000 hectares (more than 40 times the area of BASF's Ludwigshafen site). Many Chinese chemical park authorities have visited world-class chemical parks in other countries to further improve their new parks. The most important locations in China are the Nanjing Chemical Industry Park and the Shanghai Chemical Industry Park, so-called state level chemical parks. BASF is a major investor in both of these.

Nanjing is located about 300 kilometers up the Yangzi River from Shanghai. Large ocean tankers up to 50,000 tons (100,000 tons after 2006) can use the

Yangzi River to transport goods to and from Nanjing, which is a traditional chemical manufacturing center in China. The following chemical activities are clustered in Nanjing:

- Chemical activities of the Sinopec group (China Petroleum & Chemical Corp), an integrated oil and chemical company, organized in four companies in Nanjing: Yangzi Petrochemical (YPC), Nanjing Chemical Industry Co., Jinling Petrochemical (JPC) and Yizheng Chemical Fibre;
- Two world-scale refineries operated by Sinopec group companies (YPC and JPC): they are connected via pipelines to oilfields in Shandong province (northeastern China) and after 2005 will also be linked to the deepwater harbor in Ningbo;
- Nanjing Chemical Industry Park: founded in October 2001 as the new center for the development of the chemical industry in Nanjing. The start-up phase covers an area of 380 hectares (land preparation finished). For the start-up phase utilities and infrastructure will be provided by the Sinopec group company YPC. Concerning logistics, 22 jetties are available at the existing chemical companies (accessible for vessels from 5,000 to 25,000 tons). In addition, Nanjing Chemical Park has railway and expressway access. A variety of projects are being pursued in the Nanjing Chemical Park. One of the major projects is the Celanese acetic acid project with an annual capacity of 600,000 tons.
- Adjacent to the chemical park, BASF is implementing its largest investment in Asia: the construction of a world-scale integrated Verbund site within the legal framework of BASF-YPC Company Limited (BYC), a 50:50 joint venture of BASF and Sinopec. Total investment is US$ 2.9 billion.

The investment in Nanjing is a cornerstone in BASF's development in Asia. The project negotiations started in the mid-1990s. In 1996 the two partners signed a letter of intent and one year later the project proposal was approved by the Chinese authorities. After approval of the joint venture contract in mid-2000, BYC obtained its business license in December 2000. Commercial operation will start in mid-2005. BASF has been a pioneer in China by pursuing a world-scale Verbund project at an early stage.

The heart of our Verbund structure in Nanjing is a naphtha cracker with an annual capacity of 600,000 tons (*Fig. 14.6*). It produces ethylene as the starting material for the C2 value chain, propylene as the starting material for the C3 value chain and pygas as the starting material for the aromatics value chain. Ethylene will be used to produce polyethylene, ethylene glycol and propionic acid. Propylene will be transformed to yield acrylic acid and C4 oxo alcohols. As in Kuantan, both products will be used to produce acrylates. Oxo alcohols will also be sold directly to the Chinese market. Using pygas, the aromatic extraction unit will yield benzene, toluene and xylenes.

Since 2004, natural gas has been available in Nanjing from the West-East Gas Pipeline, which brings natural gas from the resource-rich west of China to the industrial areas of eastern China. Natural gas is the starting material for the production of syngas and hydrogen in BYC and the feed for a cogen power plant. Within the Verbund structures of BYC, syngas and hydrogen are used for the synthesis of propionic acid, C4 oxo alcohols, formic acid and DMF. Dimethylamine, the other raw material for DMF, is produced from purchased ammonia and methanol. Mono- and trimethylamine will be sold to the market.

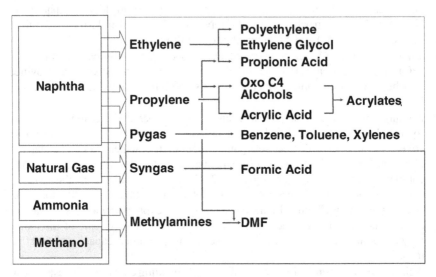

Fig. 14.6. Product Verbund at Nanjing (simplified)

Nanjing is an integrated Verbund site comprising the C1 (natural gas-based), C2, C3 and C6 value chains. There will be a balanced energy Verbund between steam producers and consumers. BYC will operate its own cogeneration plant with three gas turbines and one steam condensation turbine. Concerning logistics, BYC will operate four jetties of its own and will be linked to the national railway and expressway systems.

The construction of the integrated chemical site in Nanjing is a monumental task. Some ball-park figures illustrate this. The design of the project took 3.5 million hours – one person working eight hours a day would need 1,200 years to complete this task without being able to take a single day off. At the peak time of construction, about 14,000 workers were working on the site. Just for comparison: during construction of the Empire State Building in New York, a maximum of 4,000 people were working on the site. Obviously it is a tremendous task to organize and supervise such an enormous number of people. The Verbund site will contain 1,100 kilometers of pipes and 350,000 cubic meters of concrete will be used in its construction (1.5 times the amount used for the Empire State Building). 150,000 tons of steel will be used - sufficient for the construction of about 20 Eiffel towers. In a nutshell: the construction of our integrated chemical site in Nanjing is a tremendous task and challenge. Its completion will be a major step ahead for BASF's development in Asia-Pacific.

14.4.3 BASF's Investments in the Shanghai Chemical Industry Park (SCIP)

A second world-scale chemical park is being developed at Caojing, 60 kilometers south of downtown Shanghai. The Shanghai Chemical Industry Park (SCIP) is set to rival world-class chemical parks such as Jurong Island in Singapore. It covers 2,300 hectares, about three times the size of BASF's Ludwigshafen site. The park is being constructed on reclaimed land. As it could not rely on existing infrastructure, its own world-class infrastructure has been developed.

The Shanghai Chemical Park has attracted a variety of investors. BP is collaborating with Shanghai Petrochemical, a Sinopec affiliate, in a world-scale integrated chemical complex (a naphtha cracker with an annual capacity of 900,000 tons and downstream plants). Bayer is pursuing several projects at SCIP, as are many other Western companies, e.g. Lucite, Vopak, Air Liquide and Praxair. In addition, many Chinese companies have selected SCIP for their investment. These include Shanghai Gaoqiao, Hua Yi and Shanghai Tian Yuan (Group) Corp.

BASF is pursuing two investment projects at the site with a plant for the production of polytetrahydrofuran (PolyTHF®) and tetrahydrofuran (THF) and an isocyanates complex.

In summer 2003, we laid the foundation stone for integrated production plants for PolyTHF® and THF, which are scheduled to start operations in early 2005. With PolyTHF®, a starting material for spandex fibers, we aim to become the preferred supplier to China's rapidly growing textile industry. The plant will be the first to use a new technology developed by BASF that eliminates synthesis steps, thus saving both costs and energy. The project will be implemented within the legal framework of a wholly owned BASF subsidiary, BASF Chemicals Company Limited.

Also in 2003, BASF started constructing a production complex for isocyanates at Caojing, which we will operate with Chinese partners and the U.S. company Huntsman. We intend to supply other companies - including Bayer - from these plants.

14.5 Cultivating the Best Team in Asia-Pacific

Concrete, steel, machines and money alone will not be sufficient to shape our new Verbund sites in Asia. The main driver for our success in the region is the people who working for the company there.

BASF employs close to 10,000 people across Asia-Pacific. They form the human backbone of the company and another dimension of our business Verbund: a highly effective and efficient know-how Verbund integrated also with our customers. Developing human potential for our operations in Asia-Pacific is a key task for success in the region. BASF is striving to cultivate the best team in Asia-Pacific, mainly based on local talents.

15 Bayer – A Multinational Committed to China

Elmar Stachels: Bayer (China) Ltd., Beijing, P.R. China

Bayer is a group of research-oriented companies with business operations across a range of industries worldwide. Organized under the umbrella of Bayer AG, the Germany-based management holding company consists of three subgroups. Bayer HealthCare develops and manufactures products for preventing, diagnosing and treating diseases in people and animals. Bayer MaterialScience manufactures high-performance materials used in the automotive, construction, electronics and sporting equipment industries, among others. Bayer CropScience is market leader in innovative products to protect agricultural crops.

Bayer's corporate culture has always been geared toward research and invention, which form the basis of the company's current portfolio of nearly 10,000 products. By combining a focus on research with an unsurpassed knowledge of people, animals, plants and materials, we can tap the potential of the fast-growing Chinese market. Our goals are to grow with China, to continue using our knowledge and expertise to China's benefit, and to increase our already substantial investment there as the market develops.

15.1 Bayer in China

Bayer's decision to enter China was not simply a question of assets and market share. Rather, it was a strategic move to enter a large and potentially lucrative market that can benefit from expertise such as Bayer's in the fields of pharmaceuticals, crop protection, and advanced innovative materials. These sectors will continue to be central to China's development – any modern society is inconceivable without them.

Bayer's involvement in China dates back to the latter part of the nineteenth century, when the company first sold dyes there. Production of Aspirin® began in 1936 in Shanghai. Over the years, progressively greater economic reforms encouraged Bayer to expand its business, and in 1994 we established a holding company, Bayer (China) Ltd., from which we quickly built up a substantial production base. We started – in accordance and in correlation with China's industrial development – with local production for light industry such as leather, textile and paper, followed by local manufacturing of raw materials for the fast-developing automotive, electronic and construction industries. In line with the increasing attention on public health and agriculture, we are now strengthening our activities in healthcare for the prevention, diagnosis and treatment of chronic diseases and in crop science for healthy nutrition.

Today, Greater China constitutes Bayer's second largest market in Asia Pacific, accounting for roughly one quarter of the company's regional sales. Our production facilities efficiently supply the mainland as well as regional and global markets, broaden China's access to export markets such as Japan, and provide local employees with quality jobs in the high-tech sector.

China's rapid growth has made it a primary center for Bayer's global investment activities. Current investment strategy focuses on a world-scale, world-class integrated chemical site in the Shanghai Chemical Industry Park for the global market. This project, of which US$ 1.8 billion are already committed, represents an historic investment for Bayer and ranks among the largest investments in China by any multinational company.

15.2 Bayer's Core Competencies

15.2.1 HealthCare

One of the world's largest suppliers of healthcare products, Bayer HealthCare provides effective solutions in more than 100 countries. In China, it comprises four divisions:

- Animal Health manufactures and distributes veterinary medicines, vaccines, and grooming products for livestock and pets.
- Consumer Care supplies over-the-counter (OTC) drugs and dietary supplements.
- Diagnostics develops systems for critical care analysis, diabetes, hematology, urine chemistry, immunology, clinical chemistry and nucleic acid analysis.
- Pharmaceuticals develops drug products to treat life-threatening conditions and disorders that impair life quality and expectancy.

China's pharmaceutical industry was opened to outside markets earlier than many other industrial sectors, and today nearly all of the leading multinational drug manufacturers have a presence in the PRC. Many have formed joint ventures with Chinese companies. Roughly 40 percent of all Chinese drug firms have collaborative venture projects with foreign companies, largely because of the restrictions China has historically imposed on direct imports of foreign drugs.

China's pharmaceutical market is dominated by the hospital sector, which accounts for 90 percent of all pharmaceutical sales; the remaining 10 percent reaches consumers through retail pharmacies and local clinics. More than 17,000 distributors operating nationwide supply medicines to hospitals, retail pharmacies and stores. Although rules are scheduled to change with China's WTO entry, foreign firms currently must use these distributors to get their product into the market. Success therefore depends on generating a distributor-driven push by offering more lucrative margins or on creating greater direct consumer demand through advertising and other marketing efforts.

Over the next decade, we expect the enforcement of intellectual property rights (IPR) to grow more stringent. At the same time, drug regulations will become clearer, distribution systems will grow more efficient and the country's medical needs will develop further. Together, these factors will undoubtedly increase China's attraction for drug companies and heighten market competition. They will also make business success more likely for foreign drug companies.

The consistent opening and reform process in China has encouraged us to expand our business accordingly. Bayer Healthcare Company Ltd. set up its own plant in the Beijing Development Area in 1995. The state-of-the-art facility fulfils international GMP standards, producing solid and semi-solid pharmaceuticals. One of the top ten pharmaceutical companies in the Chinese market, we have leadership positions with Glucobay®, number one in the oral anti-diabetic market, Adalat®, number three in the antihypertensive market and Nimotop®, number one in the CNS market. In the OTC sector, we are well positioned with brand names such as Talcid® and Canesten®.

Bayer's Healthcare business focuses on offering solutions to the public health issues facing China today. We address these issues in part by continuously developing treatments for emerging diseases that threaten the quality of life for millions of Chinese. Diabetes, for example, has been comparatively rare in China over the past 30 years. But today, changes in diet and lifestyle have caused a sharp increase in its incidence. Children are particularly at risk, with obesity and overweight causing more of them to develop Type 2 diabetes – a condition historically found mainly in adults. High blood pressure is also on the rise, for similar reasons. Bayer products such as Glucobay for diabetes and the antihypertensive Adalat help enhance quality of life in China.

15.2.2 CropScience

Bayer CropScience is a leading supplier to the global agricultural market, including China. Our outstanding product pipeline and the acquisition of Aventis CropScience in 2002 will enable the company to achieve sustained growth and leadership in key segments of the market in the years ahead.

Approximately 700 million farmers need crop protection products in China. Bayer CropScience has a long history in China since the first products were introduced there in the 1950s. Today, active ingredients invented by Bayer have been widely registered and introduced into the Chinese agriculture market and over 30% of all Chinese crop protection product sales are derived from products first invented in Bayer's worldwide laboratories.

Bayer CropScience research is fully committed to introducing new technology to worldwide agriculture and China. Bayer introduced its most modern chemicals to China at an early stage, and has established production facilities for its key products in China.

Chinese authorities are now actively taking steps to modernize the agricultural chemical industry by replacing older products which have disadvantages such as the relatively high hazard to applicators and residue problems. The continued use of these older products costs China an estimated US$ 9 billion annually in lost export sales. Bayer CropScience is well positioned to contribute to this modernization. In corporation with local companies and authorities, a range of low environmental impact and high performance products, such as Regent®, Decis®, Puma® and Admire™ have been introduced to China.

Bayer CropScience continually tries to improve the awareness on responsible product care issues amongst users and officials, both alone and in cooperation with CropLife China. The AgriCare™ philosophy has been developed by Bayer CropScience to meeting challenges now facing Chinese agriculture. AgriCare™ is a full scope responsible care program for the crop protection industry, which stresses the protection of industrial and farm workers, the environment and the end consumers.

The accession to the WTO will bring unprecedented opportunities to China's agriculture and also cause strategic readjustments. The Chinese government places a high priority on the development of Chinese agriculture and has formulated detailed guidelines, basic principles, targets, focuses of work and strategic countermeasures for agricultural development in the Tenth Five-Year Plan period (2001-2005). The emergence of new marketing-savvy local producers, meanwhile, will rationalize the market as farmers are offered increasingly less toxic and more sophisticated products. Bayer CropSience has cooperated with local authorities, partners as well as local commercial distributors to bring direct benefit to Chinese agriculture with its innovative agricultural solutions and services.

15.2.3 MaterialScience

As China's economy grows, its manufacturing and construction industries will continue expanding, creating strong demand for the technologically advanced polymer materials and high-tech plastics produced by Bayer MaterialScience.

Bayer MaterialScience serves a diverse range of customers across industry sectors such as automotive, electronics, construction, IT, and sports and leisure. Customers benefit from expertise and technology that make products safer, more cost-effective, longer lasting, and more environmentally friendly.

Bayer MaterialScience is especially committed to China, supplying the industries that drive China's growth and export performance. While doing so, Bayer places special emphasis on processes that reduce environmental impact, for instance eliminating the use of harmful solvents in surface coatings.

Evidence of Bayer's commitment to the market can be seen in the joint ventures we have created. These include Bayer Guangyi Panel Co. Ltd., originally set up in Beijing in 1996 to manufacture polycarbonate sheeting and other products for the Chinese construction industry. In March 1999, Bayer opened a polyol formulating plant at Bayer Jinling Polyurethane Co. Ltd., a joint venture in Nanjing.

The most compelling evidence of Bayer's long-term commitment to China, however, is its historic investment in the integrated production site situated in the Shanghai Chemical Industry Park (SCIP). Economic growth has fueled a rapid expansion in manufacturing and construction, creating demand for high-tech polymer materials and plastics. Designed to meet this growing demand, the SCIP site brings together multiple production facilities in one location to maximize efficiency, concentrating infrastructure so each unit can focus on core production activities.

The integrated site marks an irrevocable step toward making China one of Bayer's most important manufacturing platforms for domestic, regional and global markets. China's spectacular growth derives in large part from manufacturing and production - not only the production of inexpensive items such as toys and shoes, but also of high-quality goods requiring state-of-the-art technology, such as products for the automotive, IT and telecommunications industries. To serve this expanding market efficiently, Bayer has concentrated its production facilities as close to customers as possible, thereby reducing logistical complexity and minimizing costs. By the end of the decade, the SCIP integrated site will produce high-quality materials for a range of industrial applications, including coatings, polyurethanes and polycarbonates.

15.3 A Strategy for Growth

Bayer has learned many lessons about doing business in China over the past few decades. From a hiring standpoint, it is clear that multinational companies must build on the right mix of local and international human resources, supporting operations with world-class specialists and marketers. Future contenders must staff up now to gain scientific and marketing expertise in all business areas critical to China. Moreover, they must hire more systematically and train their people more intensively. Relying on a handful of expatriates or on cheaper but less qualified local staff exposes companies to the risk that a strategic core of experts could be lured away or that staff may lack the expertise needed to do their jobs properly.

As an innovative, research-based and technology-driven group of companies, Bayer has concluded cooperation agreements with universities and scientific institutions, enhancing the development of new products and processes and offering recruitment of capable local employees.

Much has been said about the importance of guanxi, translated literally as "relations" but best understood as "connections." Many people believe success in China rests entirely on guanxi but that idea is misplaced. Although relationships are important, they do not constitute a panacea or a guarantee of success.

First, guanxi cannot compensate for poor performance. Competence, quality and professionalism remain necessary ingredients for success. Second, in a market as complex and fragmented as China's, guanxi inevitably runs into geographical limitations. Relationships in one province, region or country may prove useless when the job moves beyond one's immediate backyard. Third, guanxi is a finite commodity that can be over-used, used unwisely or exhausted. And finally, it comes with a time limit, normally reached when contacts retire, switch jobs or become less influential.

For foreign companies, the importance of cultivating guanxi is surpassed by the need to become part of the communities in which they operate. In China, Bayer meets this need through its involvement in a diverse range of educational initiatives, training programs, employee volunteer projects and charitable undertakings in areas including environmental protection, poverty alleviation, education and microfinance.

Such activities, typically referred to nowadays as corporate social responsibility or CSR, have become fashionable of late. But Bayer has always recognized its responsibility to give something back to the community and stakeholders in addition to customer orientation, particularly when operating in a foreign country. The company understands that maintaining a strong sense of social responsibility enhances our standing and therefore our value. As a consequence, we view CSR as being a natural complement of commercial success, an integral component of our overall business strategy. Today, Bayer supports hundreds of social and environmental projects worldwide and maintains a number of ongoing CSR programs in China. Amongst these are:

- Public health: Public health is a major concern for Bayer. Cardiovascular disease is on the rise, cases of diabetes and high blood pressure are increasing rapidly. In addition, HIV/AIDS is one of the most serious problems facing China's population in the coming years. Bayer is addressing this situation by maintaining China-based programs to raise community awareness of HIV/AIDS. The Tsinghua-Bayer Public Health and HIV/AIDS Media Studies Program will raise awareness of public health issues by training Chinese journalists to cover them accurately and conscientiously, at one of China's top institutions of higher learning.

- Employee health and safety: Bayer is of course most concerned for its own employees, maintaining strict environmental health and safety standards at all of its production facilities. Bayer regularly assesses and audits facilities worldwide, minimizing the potential for work-related hazards. In China, a new wellness and health check program for employees will further this commitment by educating employees on a wide range of health issues.

- Special Olympics: Bayer supports the Special Olympics, an international organization dedicated to empowering individuals with intellectual disabilities. We actively promote special athletes throughout China and support three special schools in Beijing and Shanghai. In addition to funding Special Olympics activities, Bayer sends volunteer employees to lend support and participate in sporting events such as bowling competitions. Bayer's commitment in this area extends to its own hiring practices. The company's Beijing office has employed an intern belonging to the Special Olympic organization to work part-time as a liaison between Bayer and the Special Olympics activities in China.

- Environmental Envoys: Bayer launched the Young Environmental Envoys program, an educational initiative aimed at enhancing environmental awareness among young people across Asia Pacific, in cooperation with the United Nations Environment Programme (UNEP). The program brings together students aged 18 to 24, government officials and Bayer employees. Participants visit ecologically important sites and complete environmental research assignments, participating in 'Eco-Camps'. A select group are then given the opportunity to visit Bayer's facilities in Germany, where they learn more about environmental protection and waste management.

- Microfinance: Most rural poor have great difficulty financing an enterprise. Poverty limits their ability to save money and they lack the collateral banks require to receive a loan. To address this problem, Bayer China formed a microfinance partnership with several organizations through the international NGO Mercy Corps to fund a community development project in rural Fujian province. The project targets poor fishing and farming households in remote mountain areas. Partners contribute money that is used as seed capital for loans, which borrowers use to start small family businesses. Women in particular are encouraged to participate.

Bayer considers its relationship with China to be a symbiosis: a mutually beneficial partnership that allows both entities to grow and thrive. We expect China to grow and thrive, both now and for many years to come. GDP growth continues to be strong and last year imports grew 40 percent, making China an engine for the rest of Asia Pacific.

But while continued growth is vital to prosperity, Bayer is careful to focus on achieving it in a way that is profitable, healthy and sustainable. The challenges involved in doing this are indeed numerous, but we look forward to mastering them, together with our Chinese partners.

15.4 Summary and Outlook

Bayer's investment policy follows the principle of investment in accordance and correlation with China's industrial development and local needs: first investing in local manufacturing for the light industry such as textiles, leather and paper, followed by investments in the production of raw materials for key industries such as automotive, electronic and construction. Based on the track record of a consistent step-by-step investment approach, Bayer is investing into an integrated site as a platform for supplying key industries in the domestic market as well as for exports. In line with the increasing attention on public health and agriculture, Bayer is now strengthening its activities in healthcare for the prevention, diagnosis and treatment of chronic diseases, in crop science for healthy nutrition and in technological services for environmental issues.

The holding company has been developed as competence center to

- facilitate the various investments in joint ventures and WOFE's since 1994,
- ensure group quality standards and processes,
- focus on win-win projects, which contribute to the development of the host country, and

- enhance CSR (corporate social responsibility) activities, such as training and education and community projects.

After the strong focus on dynamic growth the country is now emphasizing a balanced and sustainable development between economy and society with more organic and consistent growth. Issues regarding healthcare, social welfare and rural income and development are high on the government's agenda.

Bayer is prepared to contribute to these long-term goals with its business operations and its numerous activities. The overall policies fit in Bayer's new strategic alignment, concentrating on the R&D and technology based, innovative driven business areas Healthcare, CropScience and MaterialScience.

Enhanced life quality and increasing education levels paired with growing information and communication technology promise further growth perspectives.

16 Bicoll – The First Sino-German Biotechnology Company

Kai Lamottke, Nicole Feling and Christian Haug: Bicoll Biotechnology (Shanghai) Co. Ltd., Shanghai, P.R. China

Bicoll is the first Sino-German biopharmaceutical enterprise. It was founded in 2001 with two legal entities: Bicoll GmbH in Munich, Germany, and Bicoll Biotechnology (Shanghai) Co. Ltd. in Zhangjiang Hi-Tech Park, P.R. China. Technology development, cooperation management and marketing are functions in Munich, while facilities for research and development are located in Shanghai. The company is specialized in high-tech natural product chemistry and drug discovery services. The focus is on making compounds from natural resources compatible with customers' (drug) discovery systems in Europe and North America.

16.1 Considerations for a Chinese Site

16.1.1 General Business Environment at the Start

After the biotechnology boom in the late 1990s, founding a biotechnology company in Germany in the year 2000 was already difficult (*Arnold/Lamottke* 2003). The first signs of general recession were evident, the mood of the market for biotechnology companies became increasingly changeable and the downhill trend for the whole line of business and financing began. The deep pockets of a large organization, which helps to pre-finance the first steps or even serves as back-up resources in the event of management mistakes, were not available at the start in 2001. It was even harder to promote the idea of setting up a parallel enterprise in Europe and in a developing country like China.

And what about the business environment for Bicoll in China in 2000? The country's economy was booming and its chemical industry was playing an increasingly important role in the global markets. Nevertheless, the financial instruments for start-up companies were and remain limited. At the point of Bicoll's foundation in China, no examples existed as guidance for the successful establishment of a private foreign-funded, high-tech laboratory in the field of biotechnology. Therefore, a novel biotechnology business model tailored for the parallel set-up in Europe and in China had to address this limitation. An early positive cash-flow was one of the most important milestones in this innovative concept.

16.1.2 Underlying Business Concept for Parallel Set-Up

The decision to set up Bicoll parallel in Munich and Shanghai despite the challenging business environment was motivated by the initial business idea developed in the mid-1990s: how can the most successful sources of new and innovative drugs for human well-being - natural products - be integrated into modern drug discovery?

Besides technology, the most essential question at this time was how to safeguard access to the plant materials, which are mostly found in developing countries. The 1992 Rio-Convention for the protection of biodiversity severely restricted this access for industrialized countries. In many cases, the species containing interesting compounds for drug research and development had been exploited in the past, without any regard to their preservation. As for the fauna, a vast and careless harvest had often endangered or even destroyed species. Although the Rio-Convention raised these questions, it did not deliver successful business models other than those on a non-profit, and therefore unsustainable, basis. The only way to address such an issue pro-actively was to set up a privately funded, research-driven company directly in the country of organism origin. This business concept also involved local people in many steps of the innovative process – one of the fundamental conditions of the Rio-Convention.

In times with a difficult funding and growth capital situation, the lower burn rate and operational costs in Shanghai compared to Germany also helped to start the business successfully, even if the cost savings were not as great as is often believed. In particular, if products are aimed at Western markets as in Bicoll's case, it takes a lot of additional effort and management resources to make R&D products visibly competitive in the market.

In translating this business concept into action, a solid basis was provided in that the members of the founder team already combined China and Germany by their national origin. Therefore, for Bicoll, it was not a question of finding the right partner in China as is the case for most Western enterprises entering that country – Bicoll has been a genuine Sino-German enterprise right from the start.

16.1.3 Global R&D Environment and Conclusions for a China Entry

Two trends have shaped the character of global pharmaceutical R&D: the decryption of the human genome and the automation of chemical and biological research. Since the end of the 1990s, decrypting the human genome followed by combinatorial chemistry, computational rational drug design and high-throughput screening systems seemed to be the ultimate Holy Grail for the new millennium's drug discovery process. Natural product chemistry was mostly considered old-fashioned and everybody hoped to find arguments to close such kind of development departments as soon as possible.

Prejudice and the bad reputation of natural products are most probably derived from different technological strategies in dealing with naturally occurring substances and an extremely wide variety of sources, e.g. venoms, insects, bacteria, marine organisms, plants. All of these are generally subsumed as natural products but not all strategies are successful in terms of research output and ultimate business success. This often results in an organization's conclusion that dealing with natural products is not successful and the route is then rejected as a means of delivering the starting point for development.

However, a careful analysis of existing drugs on the market yields the opposite assessment of the current situation. Around 60 percent of the nearly 900 new small molecule chemical entities introduced as drugs worldwide during the last 20 years can be traced back to natural products (*Newman/Cragg/Snader* 2003). One impressive example is the most successful and best-selling drug ever: Lipitor[1] with current annual sales of around EUR 10 billion. This product belongs to the compound class named statins which were first discovered in natural sources. After Merck successfully launched the first and second generation of statins, there were soon more than 14 companies which launched drugs in that class, including Pfizer's Lipitor (*Downton/Clark* 2003).

The high expectations in fast results from genome decryption on the one hand and poor results from drug discovery by combinatorial chemistry on the other hand have been dashed already. Thus, natural products strategies for novel drugs are worth to be evaluated again.

In China the domestic pharmaceutical research and development environment is dominated by universities and scientific institutes rather than pharmaceutical enterprises. These institutes are fast closing the gap on Western research institutes in fields where established standardized processes are applied, e.g. genome research. This is much harder in highly innovative fields with a risk of failure, such as the identification and validation of disease targets. Hence, domestic private enterprises are hesitating to research new drugs on their own. As a consequence, domestic firms hold only 3 percent of the intellectual property rights to all chemical drugs on the Chinese market (*Chen* 2004). According to the State Food and Drug Administration of China, the remaining 97 percent are legal imitations of imported drugs. Even if entry into the WTO forces many companies to strengthen their intellectual property position with innovative and patented products, one should not forget that the Chinese market for cutting-edge research and development services is still at an early stage. Therefore, unlike for most Western corporations heading to China, the opportunities offered by the Chinese market are not the main motivation for Bicoll. Bicoll's set-up in China is mostly motivated by the factors of an existing local network with high competence in the field of natural product chemistry, a novel technological approach and high demand for new drug candidates on a risk-sharing price basis.

In the current global and Chinese R&D environment, dealing with natural products successfully is not merely a scientific question but in many ways primarily an organizational and management task. Bicoll's natural product concept is based on the strategic option of fostering results throughout early drug discovery and development and of offering outside support for the initial steps of product creation. A closer look at the technology and business model will show how Bi-

---

[1] Lipitor® (atorvastatin calcium) is a brand of Pfizer Inc.

coll's concept helps to overcome the current hurdles in pharmaceutical drug discovery to reestablish natural products as the most successful source for drugs.

16.1.4 A Superior Starting Point for Modern Drug Research

In evaluating drug candidates, pharmaceutical R&D relies heavily on biologically similar organisms which reflect a certain type of disease in humans. This is scientifically underlined by the fact that even organisms separated by hundreds of million years of evolution (e.g. humans vs. fruit flies) rely on more or less the same basic principles in developing a metabolically unstable molecular flux, otherwise referred to as disease. Nobody doubts this approach (*Table 16.1*).

Table 16.1. Sequence of the signature of amino acids of the highly conservative K+ channel (single-letter code for the amino acids – identical pattern is bolded) (*MacKinnon* 2004)

Bacteria:	TA**TTVGYG**
Archaea:	TA**TTVGYG**
Plant:	TL**TTVGYG**
Fruit fly:	TM**TTVGYG**
Worm:	TM**TTVGYG**
Human:	TA**TTVGYG**
Mouse:	SM**TTVGYG**

This ultimately yields the hypothesis that, if there is any human disease that can be addressed by small molecules, it has already been used as a target by other organisms in the race for the survival of the fittest. In most cases, these targets are addressed by small molecules that help the producing organisms to survive. Plants are an incredible rich resource of unique compounds that act as chemical controls in hostile environments. An organism which influences the behavior of an enemy will certainly have a competitive advantage (e.g. sedatives, immunosuppressants, chemotoxic agents etc.). That is the reason why natural products are so successful in delivering new ideas and approaches in pharmaceutical research.

The use of natural resources, particularly plants, as therapeutics has always been part of human culture, in particular in China where even today most people rely on the individual therapeutic approach of traditional Chinese medicine. Nevertheless, only a tiny amount (less than 5 percent) of the more than 400,000 known species of plants has been systematically investigated in modern screening systems for the pharmacological activity of their compounds.

16.1.5 Addressing Current Needs in the Development Cycle

Taking advantage of the unique properties of natural plant products (e.g. solubility, drug-likeness, uniqueness etc.) "developed" by 3 billion years of evolutionary pressure in combination with Bicoll's technology to dig up this treasure is the jump-start to a modern Western approach for innovative drug development. Besides shortening development times, Bicoll addresses the essential prerequisites for promising substances in modern drug development:

- Quality (all of the delivered molecules have inherent drug-like properties);
- Quantity (highly diverse set of molecules);
- Selection (library customized for the disease target(s) – concept of *property-based screening* of natural products implemented by Bicoll in 2001);
- Efficiency (only active molecules are developed further and resources are made available solely for them);
- Complementary (matching natural products to "automated" test systems).

Bicoll's primary products are optimized small molecule libraries for drug discovery in the biopharmaceutical industry. The libraries are tested and developed in alliances. They are customized to the target protein of the partner's therapeutic field to provide a high success rate in screening and further development. The starting material is generated from endemic Chinese plant resources.

Bicoll has two proprietary core process technologies BIFRAC N and BIPRESELECT[2]. The proprietary BIFRAC N separation and isolation system is used in the rapid detection of biological active compounds. It is specially designed to be robust enough for use in a developing country like China. This process selects only high-quality active agents that have a good chance of further development into new therapeutics. Therefore, unlike other companies working in the drug discovery process, Bicoll does not waste any effort generating data on molecules that have little chance of ever being successfully developed into pharmaceuticals or ingredients in other related fields. Another advantage is the fact that this process needs only small amounts of plant material. Bicoll's technology shortens the long and expensive process from discovery to pre-clinical testing.

The BIPRESELECT pre-selection tool narrows in on compounds with selected ADME criteria (*property-based screening*). Thus, unsuccessful drug candidates which would otherwise block the drug development pipeline are eliminated early.

16.2 Starting from Scratch in Shanghai

16.2.1 Choosing the Right Partners and Site

At its research location in Shanghai, Bicoll combines the technology for drug discovery developed at its Munich site with the rich plant resources from defined Chinese regions. This concept of process technology transfer today involves local employees in many steps of the innovative process. During the founding and feasibility phases of Bicoll Biotechnology (Shanghai) Co. Ltd. as a subsidiary of the German-based Bicoll GmbH, the concept found wide acceptance among the local (Shanghai) and national government bureaus.

---

[2] BIFRAC N and BIPRESELECT are trademarks of Bicoll GmbH, Germany.

But due to the fact that Bicoll was the first German-Sino-venture in the field of pharmaceutical research established successfully in China, case studies for the founding were not available. Assisted by the Delegation of German Industry and Commerce, Bicoll was able to deal with all the formal (and informal) problems in setting up an initial infrastructure in Shanghai. The successful implementation of the concept as well as activities such as Bicoll's technology transfer from Europe to Asia, was rewarded with a public-private partnership program involving Bicoll and the German Investment and Development Company (DEG)[3].

DEG was the partner of choice because it is one of Europe's largest development finance institutions specializing in long-term project and corporate financing for private-sector investments. This cooperation resulted in the establishment of a research and development facility for 15 skilled and scientifically trained employees.

Crucial factors for choosing Shanghai and the Zhangjiang Hi-Tech Park industrial zone as the location for the research laboratory have been the superior chemical research infrastructure including skilled researchers and technicians, supply with lab consumables, solvents, and orderly chemical waste disposal. For the interior construction of the so-called incubation buildings provided by the industrial zone, Bicoll worked together with the German architecture office of C. Saak to meet the company's needs. In addition to several pharmaceutical and chemical companies, the industrial zone is home to a number of Chinese research institutes, which enables integration into the academic network. All these factors are prerequisites for producing and selling top-quality research services. In return, the comparatively high costs of Shanghai as a research base (compared to the rest of China) have to be accepted.

16.2.2 Organizational Structure Response for Bicoll's Two Locations

For Bicoll's business model with operational units in both Germany and China, a careful analysis of the weaknesses and strengths of both locations results in the functional structure shown in *Fig. 16.1*. The avoidance of double functions at an early stage is especially important for a resource-saving start-up. The organizational functions reflect the strength of each location.

Proximity to customers, access to Western R&D markets and the development of cutting-edge technologies are the core strengths of the Munich site. At the Shanghai site the analysis results in implementation of the production and research location and, of course, access to plant resources.

---

[3] DEG press release, Cologne, February 4, 2003 and iXPOS (Außenwirtschaftsportal), Bundesagentur für Außenwirtschaft (bfai): http://www.ixpos.de/nn_7850/Content/de/01Aktuelles/News/2003/DEG18.html.

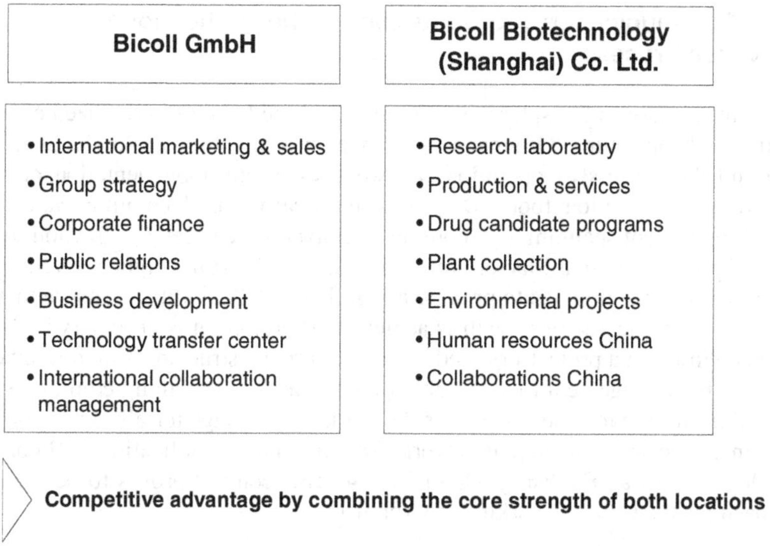

Bicoll GmbH	Bicoll Biotechnology (Shanghai) Co. Ltd.
• International marketing & sales • Group strategy • Corporate finance • Public relations • Business development • Technology transfer center • International collaboration management	• Research laboratory • Production & services • Drug candidate programs • Plant collection • Environmental projects • Human resources China • Collaborations China

Competitive advantage by combining the core strength of both locations

Fig. 16.1: Implementation plan for the allocation of local competence centers

16.2.3 Personnel Management: Attracting the Right People

Besides these rational and economic considerations, one of the most important factors in Bicoll's successful dual concept is subtitled "It's all about people". There are so many examples of how international projects fail because the partners have no open interaction or, even worse, mistrust each other. Personal relationships have been established among the scientific founders who are at home on both continents (Europe and Asia) as a result of their joint research activities long before the actual establishment of Bicoll. Hence, cultural barriers can mostly be avoided. Moreover, the long established personal contacts have a beneficial impact on understanding the two cultures.

All of the key persons involved in the foundation process also have experience in industry and of working abroad for a number of years.

Early on, a creative atmosphere is most important for additional colleagues educated in the Chinese system, which differs significantly from the European one (*Haug/Lamottke* 2004). The Chinese education system is based on teaching theoretical knowledge with little emphasis on practical application, in particular of modern methods. Therefore, it remains a primary task for management to encourage scientific employees to follow new paths, be innovative and not be afraid of failure. It is also necessary to establish the understanding that criticism is necessary to creativity and a valuable tool for the development of individual skills. This atmosphere helps employees to identify better with the company and to control fluctuation of employees. However, it also requires additional management effort.

16.2.4 Communication and Controlling – The Ignition for an Innovative Process

The communicative atmosphere is supported and guided by standardized communication and controlling. Of great significance for a smooth workflow, a motivated team and the success of operations at two sites is the implementation of well-defined communication tools. Bicoll's scientifically based communication and controlling system spanning two continents converts the necessary freedom in the research process into hard factors for delivering results at the highest level. In addition, the organization has to provide a high level of flexibility to ensure that the envisioned high degree of growth is achieved. Therefore, it is necessary to assign responsibilities on a project-oriented basis. The project structure is monitored continuously by management to avoid excessive hierarchy, something employees will demand from time to time. English is the working language for all aspects of both daily and, especially, documented work. The minimum qualification at Bicoll is a BSc degree with an English grade of CET 4. This concept proves to be the right approach for the Chinese working environment.

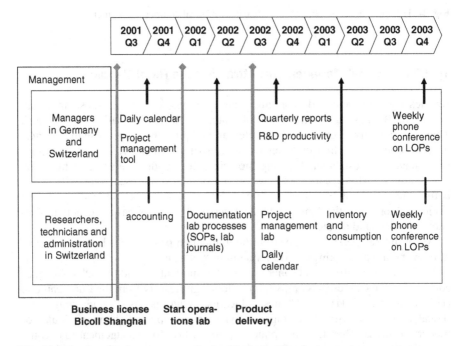

Fig. 16.2: Implementation sequence of scientific controlling instruments

Critical processes are fully documented from an early stage. It is thus possible to transfer them internally so as to best leverage in-house competence. The management team at Bicoll, Germany, is also highly involved in most processes in China so that IP-critical knowledge can be separated between the two locations and the necessary information flow can be controlled in a very effective way. For the early project work demanded, electronic process documentation and e-mail exchange are necessary to achieve symmetrical delivery of critical information. The implementation sequence of these instruments is shown in *Fig. 16.2.*

Every employee at Bicoll in both Germany and China bears a high degree of responsibility. All key personnel are involved in customer-related projects. This kind of culture fosters the understanding that at the end of a process there is a product that somebody else is willing to pay for. Progress is tracked by valid and critical check results (*Fig. 16.3*).

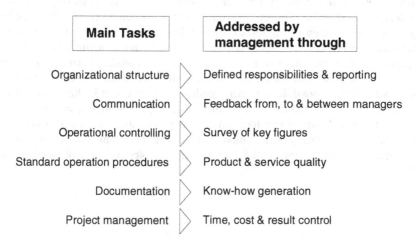

Fig. 16.3. Controlling a research-based company on two continents

As a result, all of the achieved innovations are already applied in the research processes of Bicoll and its collaboration partners. Closely allied to these processes is international project management. It is not enough to dispatch products to customers and then never follow them up. Bicoll assigns project managers with excellent skills in English, in addition to their scientific education, to support customers in working with the products supplied. The people that are able do this immediately are in fact very rare in China.

16.3 Outlook

China is indisputably a pharmaceutical market with huge growth potential. In recent years, the developed regions of the country have also made every effort to become attractive locations for high-tech R&D. In areas such as IT sector, the development gap has been closed.

Looking back at the Western pharmaceuticals market in the late 1990s, it could be observed that even big pharmaceutical companies founded their own "biotech" entities to stay innovative in this field. Since then, most of them have discovered that entrepreneurial culture and innovation cannot be created in the test tube - you need entrepreneurs themselves. As a consequence, strategic partnership is seen by such organizations as an opportunity to sustain and even foster the established entrepreneurial culture. Examples include Celera, which has bought Tularik, or Roche, which has taken shares in Genentech.

Observing the pharmaceutical industry, it appears to be fashionable to do "something" in China. Viewed from the outside, most of the research and development centers that are to be established in China seem to be more politically motivated than resulting from a unique selling preposition in their own large organization. The question remains whether the outsourcing and partnering approach that has occurred in the Western hemisphere is not the better organizational fit for big enterprises for their move into China, too. In addition to the art of linking multinational research centers, which is particularly not easy to achieve between the Western and Chinese cultures, it must be carefully considered what a company aims to do and achieve in China. Experienced and reliable partners help get a feeling for what will be possible in pharmaceutical R&D in China in the future. Bicoll thus sees itself positioned as a center with a high degree of unique competence to leverage the most from its two locations for its partnerships and its own in-house research. This has already been demonstrated in collaborations with leading European and U.S. companies in the biotech, crop protection and pharmaceutical fields.

17 Ciba Specialty Chemicals in China – Global Direction and Local Expertise

Kuno Kohler: Ciba Specialty Chemicals (China) Ltd., Shanghai, P.R. China

17.1 History

The roots of Ciba Specialty Chemicals' China operations date back to the late 19th century, when Geigy's dyestuffs were formally introduced in China by a Shanghai agent in 1886. In 1913, Ciba entered the Mainland Chinese market represented by a Hong Kong-based trading company. For the two Swiss chemicals companies Ciba and Geigy – predecessors of Ciba Specialty Chemicals – China represented the future, posing tremendous challenges and yet offering vast potential.

Ciba's and Geigy's business in China remained strong in the first decades of the 20th century. In 1935, Ciba opened an office in Shanghai under the management of Walter Naef, who had been building Ciba's business in South China since 1922. With the official establishment of Geigy Trading & Marketing Co., Ltd., China Branch in Shanghai and Hong Kong, Jardine Matheson was contracted to act as Geigy's representative in both China and Hong Kong on October 1, 1946.

By the end of the decade, however, changes in China's political climate signaled the end of a vibrant and important period of growth. By 1947, Ciba had reestablished operations in Hong Kong and in 1955 Geigy's Shanghai office officially ceased operations. It would be almost three decades before Ciba and Geigy, by then merged into one company, would reestablish formal ties with China.

As the 1970s advanced, it soon became evident that China was set to once again become an important strategic market. The Kwai Chung technical service building in Hong Kong was inaugurated in 1975. Ciba-Geigy devoted much of the next decade to gradually reopening opportunities on the mainland. Liaison offices were established in Shanghai and Beijing in 1984 and 1986, respectively.

About one hundred years after the company first began importing dyestuffs to China, it broadened its investments and presence here in the areas of additives, colors for inks, paints and plastics, textile dyes and consumer care chemicals. The group's first know-how transfer agreement, signed in 1983, was followed in rapid succession by a series of technology transfer and manufacturing joint venture agreements.

In March 1996, Ciba-Geigy and Sandoz announced plans to merge and form one of the world's largest life science groups – Novartis. Following this, Ciba-Geigy's specialty chemicals divisions were spun off on March 13, 1997 and listed on the Swiss stock exchange as an independent new enterprise: Ciba Specialty Chemicals Inc.

Later that same year, Ciba Specialty Chemicals (China) Limited, headquartered in Beijing, Ciba Specialty Chemicals (Hong Kong) Limited and Ciba Specialty Chemicals (Taiwan) were set up. A number of local manufacturing joint venture companies were subsequently also established.

In 2001, a new China operational head office was established in Shanghai as a single organizational structure, with common systems, processes, a single corporate identity and a distinct Ciba Specialty Chemicals culture and public image.

Shanghai became the headquarters of Region Asia North, formed in 2004 to include P.R. of China, Taiwan, Japan and Korea. In addition, Ciba Specialty Chemicals acquired Raisio Chemicals in June 2004, creating a top-tier chemical supplier to the paper industry. The subsidiaries of Raisio Chemicals in China were incorporated into Ciba Specialty Chemicals (China) Ltd.

The current presence of Ciba Specialty Chemicals in Region China consists of a holding company in Shanghai, three trading companies in Shanghai, Hong Kong and Taiwan, six branch offices in Beijing, Guangzhou, Hangzhou, Xiamen, Kunming and Chengdu, and 11 production sites in Shanghai, Qingdao, Panyu, Guangzhou, Shekou, Zhenjiang, Xiangtang, Shouguang, Suzhou and Kaoshiung (*Fig. 17.1*).

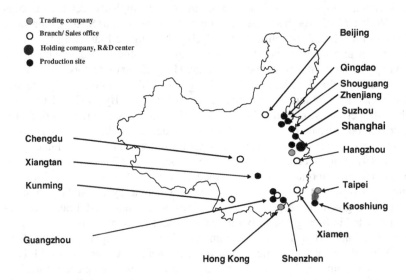

Fig. 17.1. Current presence of Ciba Specialty Chemicals in China

17.2 Industries We Serve

We have an impact that improves the quality of life – adding performance, protection, color and strength to textiles, plastics, paper, automobiles, buildings, home and personal care products and much more.

17.2.1 Plastic Additives

Additives for plastics are sold to two main customer categories. Polymer producers are usually large petrochemical companies, which manufacture plastics such as polypropylene, polyethylene, polystyrenes, ABS and a variety of engineering polymers. The plants are very often part of the downstream section of large naphtha crackers. The Ciba Specialty Chemicals' products used by these companies are antioxidants, which protect polymers against oxidation by air. They slow the ageing process, thereby extending the life span of plastic articles. Some polymers such as polypropylene are not stable in air and therefore require stabilization with chemicals to prevent premature oxidation. The number of producers of basic plastic materials in China is growing rapidly. Most Chinese producers are subsidiaries of Sinopec or CNPC. In recent years big international oil companies have started to invest in crackers in China. Examples are BP in Shanghai, BASF in Nanjing and Shell in Hui Zhou.

Plastic processors and converters operate plants that produce a variety of plastic articles for daily use. Their products range from agricultural films to woven tapes, plastic fibers, bags and injection-molded articles such as stadium seats and bottle crates. In addition to antioxidants, these plastic processors require other types of stabilizers such as light stabilizers, optical brighteners, anti-fogging and antistatic agents.

The usage of plastic materials in China is forecast to increase at an annual rate of approximately 25 percent for the next ten years. This means that the consumption of plastic materials in the year 2015 will be ten times of year 2004's volume. These are prospects which bode very well for the manufacturers of plastics and plastic- related chemicals.

17.2.2 Coating Effects

Thinking about China as a different market from anywhere else in the world would be entirely wrong as we do find the same characteristics there as in Europe or the United States of America. However, the market dynamics are more pronounced and competition is much more aggressive in China.

The Coating Effects Segment serves the paint, printing, plastics and electronic industries. In China, all major multinational companies are present in addition to some medium-sized and a very large number of small customers. A noticeable difference compared to the Western hemisphere is that multinational companies are not yet dominant market players. The fragmentation of the customer base also creates some challenges in terms of distribution, business approach, pricing and the vast area of the country, making it necessary to use distributors to reach all potential customers.

The Chinese market is extremely competitive because a lot of suppliers are by now well established. Over the years, they have not only improved their quality but have also become much more skilled in marketing. Many companies have come to China with the mindset of producing at low cost. However, producing in

China is not a guarantee for success. Local competitors do still benefit from the inconsistent law enforcement in environment, health and safety standards. This lack of a level playing field must be accepted as a limitation to doing business in China for the time being. On the other hand, this does not mean that innovation in China is not a key success factor. Local competitors are supported by long-term government backing and have access to resources from universities to master the rapidly changing chemical industry environment.

In summary, Ciba Specialty Chemicals approaches the Chinese market for coating effects the same way as it does any market. However, monitoring the competition because of its very high speed and flexibility is also much more critical in China. Despite the extreme price sensitivity, innovation is regarded as the key to ultimate success in China.

17.2.3 Water and Paper Treatment

Papermaking has a very long history in China, but the really big growth only started just about ten years ago when the industry started to privatize.

By the year 2004, China is home to the world's biggest papermaking machines with an annual capacity of up to 700,000 tons as well as to some medium-sized machines with an annual capacity of around 450,000 tons. Some of China's paper mills have reached record running speeds and many are equipped with the latest technology. But the Chinese paper industry still maintains quite a few rather small paper mills with papermaking machines running at a capacity of a few thousand tons per year. According to informal statistics, China has about 3,500 paper mills and consolidation to about 1,000 units is widely expected in the next few years. 2002 saw production of 37.8 million tons of paper and around 15 million tons of pulp in China. Total paper consumption was close to 43 million tons and ranked second only after the United States. However, per capita paper consumption at 37 kilograms is still ten times lower than the figures for the United States and most developed countries.

The paper industry is expected to see annual growth of around 10 to 12 percent and in the next three to four years additional papermaking capacity of some 13 to 15 million tons will be installed. This increase in capacity offers paper chemical suppliers great opportunities for growth. The encouraging news for paper chemical suppliers is not just the growing volume but also the need for improved paper quality. This trend fits especially well with the marketing approach and product development ability of Ciba Specialty Chemicals.

The acquisition of Raisio Chemicals made Ciba Specialty Chemicals' paper chemicals portfolio even wider. Now Ciba can offer additional products such as AKD sizes, rosin sizes, ASA, cationic modified starches and latex for paper coating, in addition to its strong range of retention and drainage chemicals, coloration products such as dyestuffs and pigments, color formers for imaging applications, fluorescent whitening agents, fluorochemicals for oil repellents and wastewater treatment chemicals for paper mills.

In order to exploit the growth opportunities of this huge potential for paper chemicals in China, Ciba Specialty Chemicals has established several production plants in the country to localize supply and provide speedy delivery to Chinese paper mills. These production facilities are close to the strategic paper producing locations. At Zhenjiang, Jiangsu Province, Ciba has established a new latex plant. In Suzhou, also Jiangsu Province, it has acquired from Raisio an AKD emulsion and rosin emulsion plant. In Guangzhou, Guangdong Province, a joint venture to produce fluorescent whiteners is in place. Another joint venture company producing color formers is located in Shouguang, Shandong Province.

Rapid development also gives rise to entirely new demands. One of the most important in Asia is the need to get a grip on environmental pollution. Among scarce resources, water is the most important to life but also one of the most threatened. China has more cities with over a million inhabitants than any other country, but so far operates only about 400 water treatment plants. Driven by upcoming major international events such as the 2008 Olympic Games and the 2010 World Exposition, over 300 additional wastewater treatment plants are currently under construction. Shanghai plans to treat 70 percent of its wastewater within three years. We intend to seize this opportunity to market our highly effective flocculation agents for water treatment.

Under the 10th Five-Year Plan (2000-2005), Chinese government has allocated the huge sum of US$ 30 billion to water pollution control. The objective is to raise the national wastewater treatment rate to above 60 percent in all cities by 2010, up sharply from 23 percent at present. In particular, there will be a greater focus on municipal wastewater treatment as a result of growing urbanization.

The water shortage situation is exacerbated by the severe inadequacy of water treatment facilities. The Ministry of Water Resources noted that of the 60 billion tons of wastewater generated annually, industrial wastewater accounts for 69 percent while municipal wastewater accounts for 31 percent. Only 23 percent of the total wastewater generated by industrial and urban users is treated before discharge. Hence, there is a large and expanding market for water and wastewater treatment in China. Ciba is determined to be a major player in this segment.

The Chinese government has realized that, in order for the economy to sustain GDP growth of 7.8 to 8.3 percent over the next 10 years, it is imperative to ensure water supply and wastewater treatment in line with economic growth requirements.

17.2.4 Textile Effects

As a part of Ciba Specialty Chemicals, the Textile Effects Segment is a world leader in products, processes and services for the textile industry. With over 150 years of experience, this segment provides the textile industry with full packages comprising products, processes and services: from fiber manufacture, through pretreatment, dyeing and printing to finishing and the end product.

It works closely with all players in the textile value chain and extends the traditional value chain concept of an industrial company from fabric producer, machine manufacturer and textile mill to garment maker, retailer and brand house to provide truly integrated textile solutions that leverage the segment's global network (of several manufacturing sites, 50 technical centers and a sales and technical support presence in more than 110 countries).

Ciba Specialty Chemicals' Textile Effects Segment initiated its business exploration activities in the Chinese market in the early 1980s. Subsequently, it established a dyes synthesis plant in Qingdao, that has been operational since 1998 and supplies its products to both the local and global markets. The chemicals synthesis plant at Panyu (Guangzhou province) was developed in 2000 to supply the Chinese and other Asian markets. In addition to the major synthesis sites, three formulation and distribution centers have also been established in Qingdao, Shanghai and Shenzhen to enhance proximity to customers.

To further address the importance of close customer relations, the Asia-Pacific Regional Center was established in Panyu in 2001. It incorporates a world-class regional technical center and a regional training center providing innovative solution, technical support and training to our customers in China and other Asian markets. In July 2004, the Textile Effects Asia-Pacific Regional R&D center was also opened at the same location.

A local sales and technical workforce is in place in all key textile industry provinces, including Guangdong, Zhejiang, Jiangsu, Shanghai, Shandong and Fujian. A team of international and regional experts has also been relocated to China to work together with the local workforce to serve its customers.

With our continuous strong growth in recent years, China has now become the second biggest market for Ciba Specialty Chemicals' Textile Effects Segment worldwide. With the elimination of global textile quota restrictions in January 2005, China will become even more important for the textile industry. Ciba will continue to invest and provide integrated textile solutions to meet the needs of this growing market.

17.3 Business Environment

17.3.1 Environment, Health and Safety Issues

Ciba Specialty Chemicals, like most other international companies is fully committed to implementing an environment, health and safety (EHS) standard for all our operations which is in line with our stringent requirements. With that in mind, the aim from the very beginning was to build a highly professional EHS team. Due to the growing workforce and a typically high staff turnover rate, the resources needed for ongoing training efforts are high. On the basis of recorded lost-time accidents, our operations in China have achieved a level well in line with our plants in Europe or the United States.

At the early stage of our investments, most technical equipment was imported. Thanks to the build-up of the necessary professional know-how in China, we are now able to source more technical equipment locally, without compromising on safety standards. The same is true for raw material supplies where local sourcing is a key success factor in today's business environment. The big challenge in this respect is how to consistently apply our commitment to integrate EHS issues as a key element in all sourcing activities.

In recent years, local laws and regulations have been changing very fast. China's EHS legislation is now close to or higher than the international level in some areas. Together with peer companies, Ciba Specialty Chemicals strongly supports this trend and endeavors to support local government in drafting more practical and consistent EHS laws and regulations. However, this process is still at a very early stage and Ciba faces many areas where local laws and regulations often conflict with each other, are not practical and, in many cases, are not enforced by either provincial or municipal authorities. This situation is a major obstacle for all international companies, which strive to be in full compliance with all local laws and regulations independent of local law enforcement. In particular, the extremely strict COD limits in discharged wastewater place a high burden on production plants. Here we would much prefer to see a more practical approach which takes into account the total pollutant load discharged, the removal efficiency of the wastewater treatment system and the type of operations being performed in a plant rather than the present very inflexible and purely concentration-oriented regulations. As far as waste disposal is concerned, the preference is to incinerate all hazardous wastes. For some plants, Ciba has built its own incineration capacity. For others, it has to rely on local hazardous waste incinerators approved by local authorities.

In line with Ciba Specialty Chemicals' global EHS targets, the company has initiated many activities over the past two years to improve the energy efficiency of its operations. In many areas, high potential to improve energy efficiency has been identified. This is slightly surprising since energy costs in China are very high compared to other costs and the financial incentive to increase energy efficiency is actually much higher than it is in European plants.

17.3.2 Legal Considerations

We have witnessed significant efforts by China in recent decades to establish the rule of law, yet a transparent and consistent legal system is still under development. Business practice, especially of private entrepreneurs, is another key aspect of the risks faced by multinationals. Many companies still pay little attention to legal compliance and the strong desire for personal enrichment leads to the infringement of legitimate rights (e.g. intellectual property rights). Therefore, protecting a business against legal risks in this growing market environment is a challenge for any multinational in China.

Ciba Specialty Chemicals is aware of these risks and has resolved to face them positively by adopting the following preventive measures.

In implementing a contract management system, Ciba Specialty Chemicals has found that business practice in China supports the statement of "contact is more important than contract". While Ciba does not underestimate the importance of contacts (*guanxi*), it places equal value on business contracts. A well-drafted contract serves as a clear business guideline, preventing disputes in daily operations and providing direct evidence in the event of the need to settle a dispute.

Likewise of great importance is corporate governance. A transparent business process, well-defined signing powers and use of company chops, management compliance with the company's Code of Conduct and regular internal auditing constitute our corporate governance, which is part of the Ciba Specialty Chemicals management system in China. Ciba is confident that this will enable it to perform efficiently and prevent fraud and malpractice by individual employees.

Registration and government approval are frequently a prerequisite for legal protection in China. To successfully crack down on the counterfeiting of trademarked or patented products, the application for trademark registration and patent rights is legally required. In respect of undisclosed information, legal protection requires a holder to take reasonable measures to maintain the confidentiality of its technical and commercial information. Notwithstanding the efforts that China has made to protect intellectual property rights, know-how leakage and counterfeiting remain as significant threats to multinationals. The proactive adoption of preventive measures is therefore a high priority.

Legal compliance is part of the company's culture. Ciba Specialty Chemicals is committed to complying with the applicable law and regulations of China, thereby demonstrating its regard for the rule of law and compliance in its day-to-day activities. Ciba firmly believes that legal compliance will give it a competitive advantage in the long term and ensure sustainable growth. By setting a good example in terms of legal compliance, Ciba hopes to positively influence the behavior of our business partners and competitors.

Business opportunities and legal risks coexist in China. While a certain degree of risk taking may be necessary on the basis of practical considerations, an internal policy of preventive measures is the most efficient and cost-effective form of risk management in China's business climate today.

17.4 Strategic Priorities

17.4.1 Manufacturing

In the 1990s, after China began to open its market to foreign investors, the former Ciba-Geigy was one of the first chemical companies to commit a sizable sum to investment in China. Between 1994 and 2000, the company set up a total of seven manufacturing sites, all of them as joint ventures with Chinese partners. Six of these companies are now majority-owned by Ciba Specialty Chemicals. Ciba's ownership ranges from 75 percent to 95 percent. In only one case, the company has a minority holding of 49 percent. In 2004, two more sites were added to the portfolio after the acquisition of Raisio Chemicals.

The output of most of these sites is aimed at the local market. Although they have also exported right from the start, the ultimate objective is to sell most products from these plants in the local market. By 2004, about 50 percent of production output was exported and 50 percent sold locally. This ratio will further improve in favor of local sales as time progresses.

The following products are produced at the various Ciba Specialty Chemicals sites in China:

- Shanghai Ciba Gao-Qiao Chemical Company: Antioxidants for plastics
- Qingdao Ciba Dyes Company: Disperse dyes and textile chemicals
- Qingdao Ciba Pigment Company: Conventional organic pigments
- Guangdong Ciba Specialty Chemicals: Textile Chemicals
- Xiangtan Chemicals & Pigments Co. Ltd.: Quinacridone pigments
- Guangzhou Ciba Specialty Chemicals Co. Ltd.: Optical brighteners
- Shenzhen Ciba Specialty Chemicals Co. Ltd.: Textile chemicals
- Latexia Chemicals Jiangsu: Styrene-butadiene latex
- Raisio Tianma Suzhou: AKD and rosin emulsions
- Ruikang Shouguang: Paper color formers

All Ciba Specialty Chemicals' manufacturing sites in China are part of the holding company, which provides sales, promotional, human resources, legal and accounting services to these subsidiaries. In this way, the holding company contributes to simplifying the administrative and management processes of its subsidiaries.

These investments reflect the company's firm belief in the necessity of a strong local presence to secure its long-term success in China and Asia. The demand for chemicals in China is expected to continue to increase over the coming decades and a strong presence in China - including manufacturing activities - is a precondition for future success.

17.4.2 Innovation

Many multinational companies follow the usual sequence in getting involved in China. First, they set up a sales force and go into local manufacturing. As they deepen their involvement in China, they then establish research and development capabilities.

Ciba Specialty Chemicals believes that innovation is the key to profitable and sustainable business growth. The objectives of R&D are therefore to deliver innovative effects and to ensure the future success of the business. Accordingly, Ciba Specialty Chemicals will focus its annual research budget more on new markets and technologies in an effort to increase the sales share of new products in the portfolio. Under the new innovation initiative, the company has set a mid-term target of having one third of the portfolio accounted for by products that are less than five years old.

Innovation consists of developing new products, new chemistries, new technologies and new services. Ciba Specialty Chemicals is currently setting up a new research and technology center in Shanghai. This will combine expertise in organic synthesis, physical chemistry, polymer chemistry, photochemistry and analytical chemistry, as well as in application and formulation sciences, to create new products, develop new applications and provide solutions with high-value effects for the chosen industries that we serve. The new R&D center in China will be an integral part of Ciba Specialty Chemicals' global R&D network, serving both regional and global businesses.

The decision to set up a new R&D center in China is a strategic one. Clearly on account of the market size and growth, the chemical industry in China is of strategic importance to Ciba Specialty Chemicals. Our resources must follow the market and we need to extend our innovative competence to all our strategic markets.

China also offers industrial companies a ready supply of scientists. There are a number of universities and research institutes of high standard. Each year, China produces around 3 million undergraduate degree holders and 300,000 with postgraduate degrees. More recently, about thousand Chinese have been returning to China each year after obtaining a degree from a prestigious overseas university.

The issue of intellectual property is still a concern amongst many companies. In reality, China has put in place legal protection for intellectual property and the legal environment has been improving, but risks do exist in practice due to weak enforcement of the law. It is an issue for any company involved in research to manage and minimize these risks. In Ciba Specialty Chemicals' opinion, it is critical to educate employees about the importance of intellectual property rights and make them fully aware of these. While it is essential to have protective measures in place, one should not overlook the importance of creating a working environment built on trust and loyalty.

17.4.3 People Development

Talent sourcing, development and retention are the three major HR challenges for companies in China. In the last ten years, the supply of local talent has never been able to meet the demand of the market. Although the situation has improved, employee turnover rates remain very high. There are always companies willing to offer better pay and more benefits to attract the talent required to run their new operations.

Development opportunities, an attractive remuneration package and company culture/image are the top three criteria focused on by local talents when choosing an employer. Nowadays, more and more emphasis is given to career development opportunities: What kind of training program is offered by the company? Does the company offer overseas on-the-job training? What management education system is in place? How far and how quickly can I move up? By aiming at a higher position early in their career, local staff expects financial rewards to follow quickly. In such an environment, loyalty is rare as opportunities are widely available to skilled people.

There are a number of important points that companies must follow when operating in China. First of all, priority must be given to identifying the right candidate during the selection process. Good English language skills are still hard to find but one must beware of not choosing candidates solely on the basis of their ability to communicate in English. Education and employment track records are also important but as a lot of resources are usually spent on management development activities after a person has been hired, personality traits such as attitude, values and the candidate's expectations must also be taken into consideration.

Influenced by parental perceptions, staff in China often believe that it is the company's or the managers' responsibility to look after them. Hence, staff expect to be led in terms of their development needs. Superiors must take a much more proactive approach toward coaching, target setting and performance management. Managing employees' expectations and clear communication regarding career progression are vital as the risk of training and developing the staff just for them to move to the competition is a real one. Finally, development initiatives must be clearly aligned with business needs. In this, China is no different from other markets.

17.5 Future Development

The fundamental strategy of Ciba Specialty Chemicals in China is to become an "insider" in the long term as we firmly believe that this is an essential success factor for participating in this market of phenomenal growth potential. This is also an integral part of our global strategy. Ciba has made significant investments totaling close to US$ 400 million in China since the mid 1980s.

Until recently, due to regulatory requirements, foreign investment meant establishing a joint venture with local partners. Today, Ciba Specialty Chemicals has 11 production sites in China, through both direct investment and acquisition. With the exception of one wholly owned production operation, the others are joint ventures. Typically, the local partners' main capital contribution is in the form of land use rights. Some, on the other hand, provided assets or infrastructure and sales and distribution channels in the domestic market. In return, Ciba has contributed production technology, management systems and access to global markets. Its participation in joint ventures ranges from minority to predominant majority (95 percent). However, all major decisions basically require the consensus of the partners and therefore the make-up of the board of directors is very important. The recent introduction of "company limited by share" offers more flexibility for foreign investors in majority joint ventures (equity over two thirds) in terms of board control. Of course, with the change in the law allowing wholly owned operations, total management control is now feasible and will be considered for future Ciba projects in which we see no strategic need for a partner.

Regardless of the level of equity participation, one of the key learning points for us is the importance of building a joint venture relationship with Chinese partners based on sound and sustainable strategic fit. Without this, there will be times when the objectives and priorities of the parties do not match, draining management resources and resulting in failure to achieve business targets. It is therefore worthwhile spending enough time to identify the right partner and analyze the project thoroughly to ensure a good strategic fit.

Going it alone is becoming easier given the proliferation of industrial parks across the county. Keen competition among the industrial parks and municipalities to attract investment is driving rapid improvements in infrastructure and utility services offered to the chemical industry. However, at the same time, it is driving tough competition, with some industrial parks being developed before proper government approvals have been granted. Also, local incentives, including tax breaks and discounted land value etc., are offered illegally to potential investors. For Ciba Specialty Chemicals' investment projects, approval for industrial land use from at least provincial government level is a prerequisite for site selection. In addition, questionable incentives are not accepted and are excluded from economic evaluations. Over the years, the level of involvement of central government in small to medium-sized foreign investments has been evolving, shifting more to provincial and municipal levels. Therefore, it is now more important to build a relationship with the provincial government than with Beijing, as was the case in the past.

In the mid-1980s, Ciba Specialty Chemicals' initial investment approach was to strictly follow European standards in design and plant construction. Virtually all equipment was imported, resulting in significantly higher capital expenditures and product costs than local competition. As more and more local equipment and materials become available, and in increasingly improved quality, a much higher level of localization has become possible.

The higher investment costs in the early plants provided the company with the advantage of time, i.e. first in market, and therefore a leading position in the market. Moreover, they allow relatively inexpensive incremental expansions to capture continued market growth. Over time, with reduced capital expenditures due to increased localization for sourcing of materials and equipment, Ciba Specialty Chemicals observed faster breakeven and profitability on more recent investments.

The future for Ciba Specialty Chemicals in China looks very promising as the chemical industry is expected to grow significantly for years to come. The initial focus on sales and later on the investment wave of the late 1990s have set the foundation for the company's presence in China. However, recently China has started to offer much more to multinational companies than the conventional disciplines of selling and producing. The sourcing of raw materials is now of strategic importance and acquisitions of former suppliers or alliances with them are providing alternatives to the traditional sourcing locations in the Western hemisphere. People skills are improving at a fast rate and, especially, the vast pool of science degree holders is encouraging. R&D activities close to a distinguished emerging market are now an interesting option and may provide the all-important market pull effect for future innovation. Despite the enormous challenges associated with human resources, people aspects such as how to attract, retain and develop are becoming critical.

The success of any operation in China depends on the right focus of all strategic priorities that any company decides to set. A market with the size and growth potential of China's must be managed with a solid combination of global direction and local expertise.

18 Degussa: Transforming the China Region

Eric Baden: Degussa (China) Co., Ltd., Beijing, P.R. China

18.1 Early Mover

Already in 1933, at a time when both Germany and China were experiencing a historic period of change, Degussa AG first entered China by establishing a representative office in Shanghai. World War II as well as the two Civil Wars in China interrupted trade in the 1940s and 1950s. In 1958, within a decade of the establishment of the People's Republic of China, Chemische Werke Huels AG opened its first trading office in Hong Kong, named Huels Far East Ltd., to serve the Chinese mainland. Degussa AG resumed its own representation by founding Degussa China Ltd. in Hong Kong when the Cultural Revolution was nearing its end in 1974. Degussa China maintained representative offices in the People's Republic of China until 2004, when they were replaced by the Shanghai and Guangzhou branches of Degussa (China) Co., Ltd., the holding company Degussa AG established for China in Beijing in 2002.

However, given the many restrictions on foreign investments and on conducting business prevalent in China at the time it was not until 1988 - when China's opening to the outside world, initiated by the late Chairman Deng Xiao Ping with the creation of the first four special economic zones in 1980, started to impact the investment climate for foreigners - that Shanghai Masterbuilders Co., Ltd. was established as Degussa's first production base in China.

Degussa, through its various predecessor organizations, continued to develop incrementally in China throughout the 1980s and 1990s. In all, over a dozen companies were founded in Hong Kong and the People's Republic of China by the end of the millennium. The merger between Degussa-Huels AG and SKW-Trostberg AG at the beginning of 2001 then paved the way for a consolidation of the China organization and a strategic reassessment of the China Region.

18.2 Why China, Why Now?

Since the launch of its 'opening to the outside world' policy in 1980, China has been a market of promising potential for global companies, while at the same time proving to be difficult to operate in. A host of early foreign investments failed to meet the economic expectations of the investors. This led to a period of slower direct foreign investment in China in the late 1980s and early 1990s. Degussa established its first production joint venture in China in 1988 and cautiously added investments till the end of the millennium. 2001 sales amounted to only EUR 200 million, achieved by a workforce of 700, and total investments in China were less than EUR 100 million.

As a newly born global leader in specialty chemicals, Degussa in 2001 had attained the size, the confidence and the need to take a leading role in the world's fastest-growing market for specialty chemicals, China. In addition, China's then imminent accession to the WTO held the promise for more transparency and a more level playing field for global organizations like Degussa to do business in China.

Hence, in 2002, one year after the merger of Degussa-Huels AG and SKW Trostberg AG that formed the 'new' Degussa AG, as a result of the strategic reassessment of the China market, Degussa (China) Co., Ltd. was established in Beijing. This holding company was to spearhead growth in the China Region and unite the numerous individual entities under its umbrella (*Fig. 18.1*).

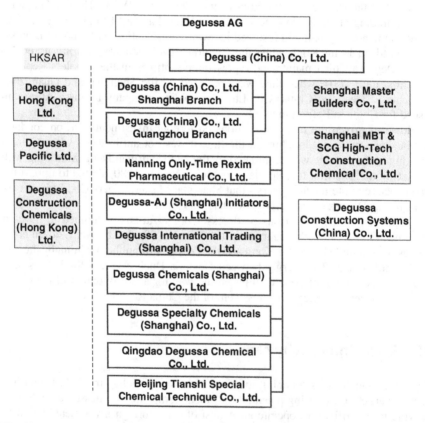

Fig. 18.1. Holding structure of Degussa entities in the China Region

China's continuously increasing strength in general, and as a producer of chemical products in particular, demands swift and determined action towards leveraging the opportunities of this buoyant market on the one hand, and towards warding off the emerging threat from Chinese competitors on the other. Before this decade is over, China will, at approximately 20 percent, be second only to Europe in terms of specialty chemicals production value, contributing as much to overall specialty chemical growth as Europe (*Fig. 18.2*). Arguably, the People's Republic of China will become the nation with the single largest market and production output for chemical products before 2010.

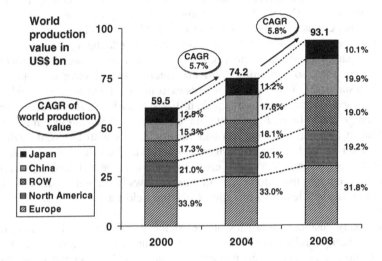

Fig. 18.2. China will account for 20 percent of the world specialty chemicals production value by 2008

The growth of the Chinese economy is driven above all by the virtually unlimited supply of low-cost, highly trainable labor. Asian economies that entered a vigorous growth period earlier due to such cost advantages, e.g. Japan in the 1970s or Taiwan in the 1980s, have seen their labor costs rise drastically and have lost their competitive advantage within 10 to 20 years. China is, however, unlikely to experience such rapid erosion of its labor cost advantage, as it has a reserve of approximately 900 million peasants who currently live on an annual income of less than EUR 300 and experience little income growth. While the living standards of the urban population rises, average labor costs in industrial production remain virtually unchanged as wave after wave of rural workers replaces urban workers at the bottom end of the labor market. Hence, despite being almost two decades into a period of vigorous growth, China is very likely to enjoy a competitive advantage from low-cost labor for another two decades.

One may argue that the chemical industry is not a labor-intensive industry, and hence low labor cost is irrelevant for the competitive position of the Chinese chemical industry. This argument neglects three important aspects. First, the level of automation in the Chinese chemical industry is far lower than in industrialized countries, where automation has largely been driven by rising labor costs. At China's level of labor costs, the productivity gains of extensive automation can in most cases not match the labor cost advantage. Hence, Chinese chemical manufacturers can command a significant cost advantage over foreign chemical manufacturers, even if both operate in China, despite a five to ten fold workforce.

Second, low labor costs have an impact on the costs of plant, equipment and materials for chemical manufacturers in China. Therefore, the savings effect of China's low labor costs is compounded if a factory is designed by Chinese engineers, built by Chinese contractors, supplied by Chinese suppliers and operated by a Chinese workforce. This mechanism, along with a strong focus on the essentials of their business, makes a large number of Chinese specialty chemicals manufacturers tremendously competitive. We are continuously updating a growing register of such companies. Particularly the thousand of small Chinese specialty chemicals manufacturers are also benefiting from lax enforcement of environmental and work safety standards. While such standards exist in China and are comparable to, in some instances even more stringent than, EU standards, local enterprises very aptly leverage the relationship orientation intrinsic to the Chinese culture and work ethic in negotiating exceptions and penalty wavers with the local environmental bureau or work safety inspectors. However, as the central government is aware of the severe long-term cost of environmental damage and occupational diseases to the economy, the enforcement of existing laws is strengthened. In the mid to long term this will lead to a more level playing field for foreign manufacturers in China.

Third, many customer industries of the chemical industry *are* labor-intensive and are increasingly relocating their capacities to China. Today, approximately 80 percent of computer screens, 50 percent of cameras and 25 percent of washing machines, for example, are made in China. The manufacturers of these goods demand the same responsiveness from their suppliers in China as they are used to from industrialized locations. The cost of providing such levels of service from a remote location, e.g. Europe or North America, fosters the competitive position of local chemical producers. Hence, as demand for chemical products shifts to China, so does their production.

Given the likely sustained competitiveness of the Chinese economy, a continued migration of our customer industries to China and an accelerating emergence of Chinese competitors on global markets, Degussa is building on its China expertise and global leadership position to leverage a strong China presence for its overall competitiveness. We are building an organization that is capable of supplying our increasing number of customers in China competitively, supplying our production sites around the world with raw materials and technical goods at lower cost, developing new customer solutions for China in China and intelligently linking our expertise and capacities around the globe.

18.3 The Holding Company – Dinosaur or Dragon?

The cornerstone of Degussa's China presence is Degussa (China) Co., Ltd. With the establishment of this holding company in 2002, Degussa became a member in the elite club of approximately 200 leading global enterprises that have been granted such rights. Degussa (China) Co., Ltd. is to support Degussa business units with a lean and scaleable service platform offering the competencies required for the Chinese business environment. Above and beyond this, the holding company is to perform strategic corporate tasks such as enforcing corporate governance, determining the direction of research and development in China, increasing contacts with China's leading universities, identifying and pursuing M&A targets, sourcing, maintaining vital networks and fostering emerging Chinese managers who demonstrate the potential for an international Degussa career.

China's accession to the WTO has fuelled a debate on whether holding companies have become obsolete. It is accurate that, once the transition period for full compliance with the WTO accession treaties has expired, the vast majority of the precious privileges enjoyed by holding companies will become common business practice for any foreign-invested enterprise. Even today, a significant portion of the services a holding company can offer to its subsidiaries may also be provided by a consulting company, without the need for ownership and with greatly reduced requirements to fulfill at its establishment. Judging the value of a holding company by legal and fiscal aspects alone would, however, neglect the importance of influence and symbolism in the Chinese business culture. No other form of incorporation is as apt in conveying, exercising and claiming power, leadership and commitment as a holding company.

For Degussa, the holding company has proven to be effective in fostering our impact on and attractiveness to important stakeholders such as current and potential partners, authorities and associations, strategic suppliers, key customers, the scientific community, as well as current and potential employees. Through the holding company, the only legal form entitled to membership in the Executive Committee of Foreign Investment Companies (ECFIC) of the China Association of Enterprises with Foreign Investment, we are able to shape the development of issues that are important to global enterprises, such as the protection of intellectual property (*Fig. 18.3*).

Pooling effects • Services • Infrastructure • Field forces	**Corporate governance** • Code of conduct • ESHQ • Segregation of duties
Market impact • Sales market • Sourcing market • Labor market	**Employer attractiveness** • Long term commitment • Career opportunities • Security

Fig. 18.3. The unique combination of capabilities that make the holding company so versatile

Often discounted and overlooked is the significance of pooling effects and corporate governance. China's developing economy confronts foreign investors with a host of challenges that have their roots in a lack of transparency. They range from unexpected fees and taxes to regulations that cannot be complied with for lack of the corresponding institutional infrastructure. The resulting gray areas can lead to a substantial and unpredictable increase in the cost of doing business. The only remedy is to establish a deep and current in-house knowledge of the regulatory environment and to link it with incorruptible governance processes. This requires a critical mass of workforce and business volume that most individual ventures do not have. Through the holding company, this vital capability has become affordable. Furthermore, due to its umbrella nature, the holding company is very effective in acting as watchdog, continuously monitoring and pruning the administrative processes of its subsidiaries.

18.4 As Decentralized as Possible – As Centralized as Necessary

Degussa strives for a decentral organization. Business units act like companies within the company and their heads are encouraged, even expected, to act like entrepreneurs. At first glance, a holding structure would seem to clash with this type of organization. In practice, the global, decentral organization and the China holding company effectively complement each other as the China holding company provides a cost-effective way for the business units to outsource any function that is not part of its core business process and to ensure compliance with corporate guidelines. At the same time, the holding company acts on behalf of the corporate center, inasmuch as it enforces the boundaries - legal and corporate - of the entrepreneurial freedom given to the business units.

The value of the central services provided by the China holding company becomes particularly apparent when growth is pursued by way of acquisition or when new sites are to be established, as is going to be the case at Degussa for the foreseeable future. Due to the continuous presence of the respective experts such as engineers, lawyers, negotiators, analysts or purchasers, the duration of a project can be shortened considerably compared to the common approach of sending teams from abroad periodically. Apart from shortening time to market, this drastically reduces the administrative cost of the project. More importantly, however, the thorough local expertise of these resident experts leads to better deals in acquisitions and to lower investments in new plants.

18.5 Sourcing China – A Profitable Growth Market

There is probably no other function where knowledge of and adaptation to the Chinese environment yields such significant economic benefits as sourcing. While many chemical companies still fret at the potential risks of sourcing raw materials, technical goods, engineering services, IT services or R&D capacity in China, others like Degussa or DuPont have realized that the opportunity cost of not sourcing here is prohibitively high. DuPont, according to its CEO Charles Holliday, expects to save US$ 200 million merely from sourcing low-value raw materials from China. In addition, the company is sourcing half of its production engineering and design in China, as well as in India and the Philippines. Given that sales in China are less than 10 percent of global sales for most multinational chemical companies, the contribution to earnings of systematic sourcing from China can well exceed that of China sales, particularly if secondary saving effects are taken into account.

However, not every company should source in China. Being successful at sourcing in China requires a strong local presence, such as a holding company complete with engineering, environmental, health, safety and quality (ESHQ) and procurement services, that is capable of specifying the goods or services to be sourced in meticulous detail, establish and certify quality management systems at the suppliers' sites, random audit suppliers and monitor the logistics chain. Building and maintaining these capabilities is costly. Smaller enterprises do not have the sourcing volume to warrant such expenses, but global leaders like Degussa can reap substantial economic benefits, as the above example documents.

In keeping with our entrepreneurial culture, we conducted a focused pilot sourcing project in 2003. With marginal risk exposure, we were thus able to understand the challenges and identify the key success factors. With the help of our own researchers and analysts we have established a database that enables us to monitor the performance of our Chinese suppliers and find alternative sources instantly whenever a supplier fails to meet our specifications. Now the time is ripe to leverage our knowledge and organizational capabilities by progressively enlarging the volumes of raw materials, packaging materials and technical goods sourced in China.

18.6 Growth

China's growth in general, and that of its chemical industry in particular, is the prime driver of market entries or expansion of existing positions by global chemical companies. At the same time, however, voices are becoming louder that warn of a bubble economy and the damage that might occur should the bubble burst. Such worries are fuelled by the high double-digit growth rates experienced over the past three years by customer industries such as the automotive, construction, electronics, electrical goods and textile sectors. When interpreting these figures it must be taken into account, however, that the fastest-growing sectors, such as automotive, have started from a very low base and that others, such as construction, are driven by the continuous need for infrastructure development in China.

Therefore, while temporary corrections cannot be ruled out, we should look into the future with confidence with regard to China's sustained demand growth, and with awe with regard to China's growth as a major trade competitor. With an average annual GDP growth of 8 percent, the industrial sector will be growing at over 12 percent compared with the mere 2 percent of the large agricultural sector. The service sector will grow even faster. Therefore, even if a slowdown is taken into account, growth in China's specialty chemical demand is likely to be in the proximity of three times that of the industrialized world.

For Degussa in China this translates into a continuation of the growth path since 2001, fostering the acquisition of controlling shares in existing Chinese businesses. During the past three years, Degussa in China has already grown to employ 1,400 people in 18 entities. Sales expanded at an average of 15 percent annually to EUR 300 million in 2004.

Continued strong growth in China is important as a significant portion of China's growth stems from the migration of production capacities to China in our customer industries. As demand from these customers fades in the traditional markets, it must be met with an adequate supply of goods and services here in China. As a customer-oriented company, we must move with our customers (see *Fig. 18.4*). The brisk growth we are pursuing in China requires substantial investments in excess of EUR 100 million annually. As any investment projects are subject to the same economic criteria as elsewhere, our China business is profitable despite the rapid growth. The myth that one cannot grow in China and earn money at the same time is not true for Degussa.

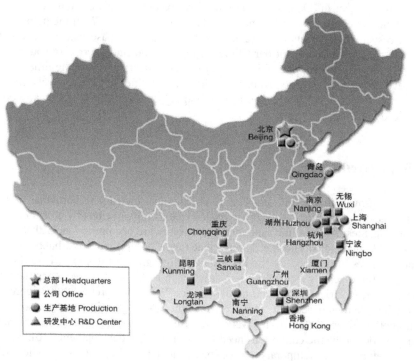

Fig. 18.4. Following our customers – major Degussa locations in China

18.7 The Other Side of the Medal

There is another side to China's two decade long economic success story. Foreign investors are well advised not to allow themselves become intoxicated by the upside potential of China but rather weigh it carefully against the downside risks present. Historically, China's industries are fragmented, with production capacities for chemicals, cars and trucks, furniture, processed food, household goods and so on in every province. This leads to the odd situation that on one hand there is oversupply for almost everything in China while on the other supplies are often more expensive than on world markets due to the inefficient production conditions. At the same time, companies face rapid price erosion for their products and increasing costs for energy, logistics and selected raw materials. Industry consolidation can and will solve this problem but is slow to happen as provinces compete with each other for the preservation of their industries and jobs. By consequence many Chinese industrial enterprises are unprofitable and in danger of bankruptcy. Credit control is an important and exceedingly difficult part of doing business with these enterprises.

Continued economic growth has strained China's infrastructure, in particular its energy supply. It will take at least two years until economic powerhouses like Shanghai reestablish sufficient power supply to support the envisaged industrial growth.

Managerial personnel too have become a scarce resource, which is expressed in soaring compensation packages for local general managers, functional heads and specialists such as lawyers. In the aftermath of the cultural revolution, there are simply too few highly educated individuals with solid experience. Recruitment, development and retention of talent are therefore rapidly developing into vital core competencies for foreign invested enterprises in China.

Even if an enterprise has succeeded in managing cost, securing supplies, building a brand and hence securing its margins, and retaining a pool of local talent to run its China operations, it must stay alert to the unpredictably changing environment. Was it feasible a decade ago to make a successful market entry into China with old technology, competitiveness increasingly depends on the ability and willingness to deploy state-of-the-art technology? This raises the specter of intellectual property rights infringement. While laws protecting intellectual property are in place, the flow of infringement cases does not cease. Penalties are still not heavy enough to deter and enforcement gives offenders ample time to abandon and rebuild elsewhere, thus avoiding execution of legal titles. Foreign investors and local entrepreneurs alike must resort to preventive measures other than the penal system. Focus is on discouraging the theft of intellectual property. Measures include black box designs, segregation of processes, and creative employee loyalty schemes. All require a rethinking of concepts that have proven themselves elsewhere.

China's WTO accession will bring down tariff protection by 2008. The central government along with the prosperous coastal provinces adheres to the agreed transition schedule, while there are numerous reports on deviations from other provinces. For the whole of China, however, the emergence of a host of non-tariff trade barriers must be expected, a practice known from other WTO members. Active participation in the lobbying efforts of industry associations is important in order to limit the detrimental effects of new legislation on trade.

Finally, the very reform process that is to ensure China's long-term competitiveness poses the gravest risk. China is walking a tightrope in its effort to keep GDP growth in a corridor of 7 to 9% while introducing a sound social security net that can support the millions of additional unemployed that will result from industry restructuring. China's success depends to a large extent on the continued influx of foreign direct investment. This in turn depends on the trust foreign enterprises have in the commitment to reform and ability to maintain social and economic stability on the part of the Chinese government.

18.8 Conclusion

China is firmly on its way to become the worlds leading specialty chemicals producer. The market opportunities in China and competitive threat from China that can already be felt demand that any firm that intends to play a leading role in the global specialty chemicals industry must secure an adequate position in China. China's characteristics favor an organizational setup that projects size, power and commitment. Success in China requires a tangible local presence, a fine meshed network of relationships and the ability and will to invest in state-of-the-art facilities. The rewards can be phenomenal when China's opportunities as a sourcing market, as R&D platform, as export base and home market are systematically and wholeheartedly exploited. China is, however, not a market for the fainthearted. The business environment is less than transparent and business practices differ substantially from those encountered in more mature markets. Cultural differences and the language barrier add to the uncertainties. Risks, of which there are plenty, are therefore very difficult to assess accurately from as far away as Europe or North America. Therefore, a strong, experienced local organization plays a pivotal role in arriving at sound management decisions regarding China. The question faced by global enterprises is no longer whether or not to engage China, the challenge has become how to capitalize on the opportunities while keeping the threats in check.

19 A Toolbox for China – Lessons from the China Experience of Degussa Construction Chemicals

Boris Gorella and Christian Kober: Degussa Construction Chemicals, Shanghai, P.R. China

> "The golden rule is that there is no golden rule."
> *George Bernard Shaw, 1903, Man and Superman*

19.1 Riding the Swift Juggernaut

China is now indisputably an economic juggernaut with a global reach. Foreign companies are naturally keen to jump onto the rapidly moving bandwagon. Many, however, are destined instead to be swept aside amid the thunder and dust of its wheels.

In this article we offer a toolbox of approaches – specific, practical advice that can reduce the likelihood of the second scenario. We have derived these approaches primarily from the hands-on learning experience of Degussa Construction Chemicals, which has operated its own production facilities in China since 1988, selling in the highly competitive construction chemicals market there. We also conducted a series of internal and external interviews inside and outside of the chemical industry to establish which of the learning experiences are applicable outside the confines of the construction industry.

What we do not attempt to offer here is a "one size fits all" guideline, because the way in which business is conducted in China has changed, and is continuing to change, very quickly, in line with the phases of political development the country has rapidly undergone since Deng Xiaoping announced the open door policy in 1978. This development has seen the country transform itself from a closed communist society to a "bourgeois dictatorship" (*Becker* 2000) in the space of just two and a half decades. So it is no surprise that quite large helpings of flexibility and adaptability need to be on the menu of any foreign company keen to establish a business in China, whatever sector it operates in.

19.1.1 The Chinese Markets

A Powerful Magnet for Manufacturers ...

Manufacturers, especially those involved in bulk materials such as steel, cement and chemicals, have been fast to identify the factors that build a compelling argument for starting production in China:

- a rapidly growing local market;
- the move of client industries to China;
- the rapid growth in the export of manufactured goods from China.

Growth in China is increasingly driven by China's internal demand. A burgeoning middle class[1] is rapidly developing the same needs and desires as its Western counterpart. Ever faster access to information has also raised the expectations of a modernizing consumer society, as reflected in retail sales which have grown annually by 16 percent since 1978 (CEIC database, 2003).

On top of this, each year millions of fresh graduates swell the ranks of China's educated labor pool,[2] with the general labor pool set to continue to grow over the next two decades. As a result, labor costs will continue to lag behind those in Europe for many years to come. Labor costs in manufacturing in the chemical industry are currently only about 30 percent of those in Europe.

The combination of these two factors, namely growing internal demand plus continuing lower labor costs, inevitably make China an extremely attractive market for most goods, provided China retains its labor cost advantage.

... and Why the Chemical Industry Is Especially Attracted

With a CAGR of 5.4 percent, Greater China is the fastest growing market for chemicals in the world, and is set to become the second largest chemicals market after the United States in coming years (*Fig. 19.1*).

Since China has focused mostly on developing basic feedstock industries in the past, the growth in specialty chemicals (*Fig. 19.2*) will be especially high, while the already established players in the feedstock industry should be able to reap the rewards of their investment.

---

[1] According to Time Magazine, 46 percent of Chinese describe themselves as „middle class", while according to the same study, only 16 percent of Chinese hold typical middle-class jobs (*Time Magazine* 2004).

[2] 2.8 million graduates entered the China labor market in 2003, while 0.5 million students graduated from German universities in the same year (*China Daily* 2003).

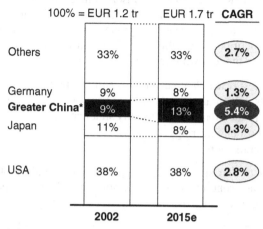

Fig. 19.1. Greater China will be the second largest market for chemicals by 2005, with an average annual growth of over 5 percent for the next decade (Source: BASF, VCI, DZ Bank, own analysis)

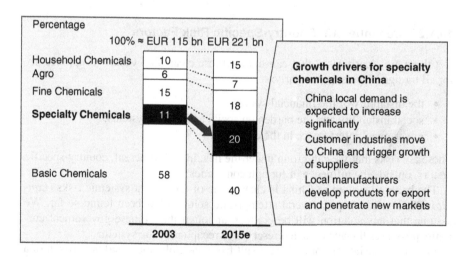

Fig. 19.2. Showing the highest growth, specialty chemicals are expected to nearly quadruple in the next 12 years in Greater China (Source: A.T. Kearney, Frost & Sullivan, Henkel, DZ Bank, own analysis)

This is why most players in the specialty chemicals field have moved aggressively in the last few years to establish themselves in China. Specialty chemicals are in an advantageous position since their products are usually not made to specification. Other segments, such as custom synthesis or building blocks suppliers, face aggressive competition from China as their Chinese competitors are in a superior cost position for the following reasons:

- access to low-cost research at universities and institutes;
- low labor costs, which are especially important in custom synthesis/building blocks, where production runs are usually small and labor intensive;
- access to low-cost engineering;
- low capital requirements.

Furthermore, products like specialty chemicals, which contain a strong service component, are not as prone to local competition, since local companies tend to lack necessary prerequisites such as:

- end-user experience;
- international R&D backup and an R&D pipeline;
- experience with value selling;
- international sales networks to leverage.

Thus, China is an especially attractive proposition for chemical specialties, as their production tends to be relatively labor-intensive, while the availability of a well-educated academic workforce also makes establishing local and regional service centers an attractive concept.

19.1.2 Three Inherent, Country-Specific Risk Factors

Yet there are clearly risks that can muddy these clear blue waters of opportunity for all foreign players, particularly:

- the instability of the financial system;
- social instability and the burden on social and health systems;
- government interference in the economy.

These are risks that we can group under the heading of "inherent, country-specific factors, outside the influence of foreign companies".

The banking system in China is clearly one of the largest systemic risks (*Langlois* 2001), and, despite several attempts, no solution has been found so far. We assume that any solution will be gradual, although the Chinese government currently possesses the accumulated reserves to recapitalize the system.

Increasing social instability may result from unemployment and the growth of a huge class of disenfranchised rural Chinese. This has been vividly highlighted by recent demonstrations in north-eastern cities. As a communist country, China is under an internal obligation to maintain a modicum of equality, and has so far managed to do so.

Further, due to China's one-child policy, the population will peak by about 2050 and will fall from then on, by which point China will have over 100 million people aged over 80 (*Jackson/Howe* 2004). The problems will become evident before the next decade is over, as China's age pyramid comes to resemble that of Western countries, with the corresponding burdens on social and health systems.

Also often underestimated is the government's continuing heavy interference in the economy, as only recently highlighted by the drastic measures taken to cool the economy. This saw the Chinese government intervening directly in the operation of steel mills, cement production and similar, dictating prices and output quantities.

However, although the challenges facing the Chinese government are formidable, China does have an enviable track record in maintaining economic and social stability, and there is little reason to doubt its ability to continue to do so. The pre-eminent focus of and justification for the continuing rule of the Chinese communist party has been the promise of stability, maintaining social order and economic growth. These policies have successfully lifted over 400 million people out of absolute poverty since 1978. We assume that the Chinese government will continue for years to come along the path of controlled growth and a gradual introduction of civil liberties.

This trend will have a significant impact on the market development of specialty chemicals, too.

19.2 Five Key Issues Facing Foreign Businesses in China

Based primarily on the experience that Degussa Construction Chemicals has gained from conducting business in China over the past 15 years and more, we have identified five key issues that face foreign businesses in China, and in the following sections present practical methods for dealing with them. These issues are:

1. Market entry, and why pragmatism and flexibility are essential here.
2. The business model, and how to avoid common pitfalls.
3. Business relations, especially trust and the true value of contracts.
4. Intellectual property – the real killer topic.
5. Human resources, primarily the problems of finding and retaining good staff.

19.2.1 Market Entry – Why Pragmatism and Flexibility Are Essential

The appropriate business models for market entry to China have changed significantly over the past 25 years, dictated both by fashion and by the government. Originally, all businesses were joint ventures with local companies, as required by law. After this law was enacted, a period followed during which most foreign companies tried to avoid joint ventures at all cost. Nowadays, we see joint ventures moving back into the limelight again, as more attractive private joint venture partners appear in the market. Furthermore, there is a growing interest in pooling the China business into holdings or similar vehicles in order to gain the same scale advantages that are available in other countries (*Fig. 19.3*).

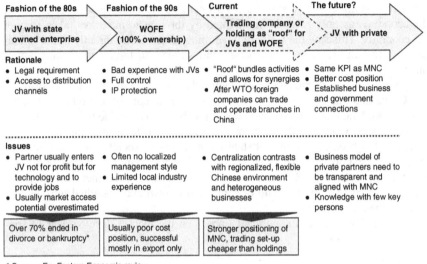

Fashion of the 80s	Fashion of the 90s	Current	The future?
JV with state owned enterprise	WOFE (100% ownership)	Trading company or holding as "roof" for JVs and WOFE	JV with private

Rationale

• Legal requirement • Access to distribution channels	• Bad experience with JVs • Full control • IP protection	• "Roof" bundles activities and allows for synergies • After WTO foreign companies can trade and operate branches in China	• Same KPI as MNC • Better cost position • Established business and government connections

Issues

• Partner usually enters JV not for profit but for technology and to provide jobs • Usually market access potential overestimated	• Often no localized management style • Limited local industry experience	• Centralization contrasts with regionalized, flexible Chinese environment and heterogeneous businesses	• Business model of private partners need to be transparent and aligned with MNC • Knowledge with few key persons
Over 70% ended in divorce or bankruptcy*	Usually poor cost position, successful mostly in export only	Stronger positioning of MNC, trading set-up cheaper than holdings	

* Source: Far Eastern Economic review

Fig. 19.3. There is no best way to enter China, but you need to be fully aware of the pros and cons of the various options

The Classical Joint Venture

Until about the mid-1990s, foreign investors had to opt for joint ventures (JVs) as their business model. This was dictated by the government, but often willingly accepted by the foreign partner due to expectations of better market access and a better cost position via the Chinese partner.

Many or most of these joint ventures failed. The core reasons for these failures can usually be attributed to widely differing expectations. Usually the state-owned Chinese partner did not actually have the market access its foreign partner expected. State-owned enterprises distributed their products by government fiat and not according to actual demand, and thus did not have a distribution structure that met the needs of the Western partner. Moreover, the Chinese partner's core reason for participating in the JV was primarily to gain both increased access to technology, as usually explicitly highlighted in the JV contract, and the chance to decrease its huge personnel overheads.

Obvious problems, like the recent release of the "Chery" car model by SAIC (the JV partner of Volkswagen AG), which is virtually an exact copy of VW's

SEAT Toledo, only serve to highlight why JVs, especially with state-owned enterprises, are largely avoided nowadays.

The Wholly Foreign-Owned Company

Due to a change in legal requirements,[3] foreign companies were allowed to set up so-called "Wholly Foreign Owned Enterprises" (WOFEs) since 1986, though in reality this became a significant possibility only from the mid-1990s onwards. Control over a WOFE is usually far easier to maintain than over a JV, no negotiations are required and intellectual property (IP) protection can be maintained more effectively. Nevertheless, WOFEs have not proved the expected panacea. Generally, their cost position suffers not only from high expatriate costs but also from a mix of other factors. New employees have to be found and are usually hired from competitors in the private or state-owned sector. They are therefore by definition more expensive compared to their peers. WOFEs were usually set up in high-cost regions (Shanghai, Yangtze delta) in order to ensure access to an international communication structure. Moreover, despite a legal commitment to exporting, many WOFEs were established with the intent of flouting this requirement. This focus on the Chinese market often resulted in investment in comparatively small facilities. These consequently do not enjoy significant scale advantages, making it difficult to compete with local manufacturers of similar or even smaller size.

The Holding Company or Similar Vehicles

As a result of the investment spree in the 1990s and owing to the insufficient scale of many operations, multinationals are trying in greater numbers to set up holding structures in China. Though the legal requirements for operating a true holding company in China are still very strict, companies have found a variety of vehicles that serve similar purposes. For example, the trade company, a new legal structure, provides most of the advantages of a holding company without the high demands on invested capital and offering a more favorable tax rate. We shall not go into more detail on these vehicles here, as holding companies are well documented investment vehicles in China.[4]

A New Trend: JV with Private Partners

As Chinese private companies have emerged, a new group of more attractive joint venture partners has come onto the scene. Such companies may be interesting partners for foreign investors in China as they share many of the same fundamentals:

- Essentially, these Chinese private companies have the same objectives as Western companies (growth, profits, securing IP).

---

[3] Wholly Foreign Owned Enterprise Law, Order No. 39, April 12. 1986.
[4] For further information see for example *Watanabe* (2002).

- They usually have a superior cost position which can be partially maintained even in a JV.
- They often have innovative know-how in respect of cost-efficient production.
- The production and engineering models of private companies often provide a welcome wake-up call to German engineering firms ("Yes, you can actually do this without having to use stainless steel!").
- They already possess their own distribution channels.
- The owner contributes his relations with the government as part of the goodwill.

Naturally, there are downsides to be considered. Typically these will include:

- Private companies tend to diversify rapidly and change business scope frequently.
- Often, their business model relies on extensive tax evasion and commission-based sales, leading to a lack of financial transparency.
- Know-how is normally limited to a few key persons.
- Environmental liabilities and land use rights are an issue in many cases.

Nevertheless, the rise of a new entrepreneurial class together with the metamorphosis of some Chinese state-owned enterprises into profit-oriented commercial ventures significantly increase the opportunities for finding a suitable partner in China.

The Need for Speed

The size and complexity of Chinese markets require sound market knowledge before actually entering the market. Yet the speed with which Chinese competition moves means that the window of opportunity between exploration and entry can be quite small indeed. As many consultants tend to be very limited in their perspectives on China, relying on the publicly available, official government database on the chemical industry as their main research tool, access to market intelligence via this route is often not possible. Consequently, a company will usually have to gain its own hands-on experience about marketing the products it intends to sell in China.

We recommend a brief, regionalized pre-marketing phase for information gathering, followed by rapid market entry. Regional pre-marketing yields a better understanding than most consulting studies can convey. Results from one region can, with the help of measures like GDP, easily be extrapolated to other areas of China. Limiting pre-marketing to a carefully selected region mostly avoids educating the competition. If good cooperation with the local government has been established beforehand, a license for importing and selling finished goods before the actual start of production may be obtained, which further increases the attractiveness of pre-marketing. A phase of three to four months can be usually considered sufficient for pre-marketing in specialty chemicals.

After this phase, the next step has to be a rapid and large-scale market entry in order to occupy the market before the competition can react. In this phase, the focus should be on capacity utilization rather than margin maximization, in order to keep the attractiveness for other entrants low. To further enhance capacity utilization, export markets should be used aggressively from the beginning. As most chemical companies tend to favor a regional approach with separate responsibilities for the China business and other Asian businesses, internal barriers to export are usually more difficult to overcome than actual trade barriers. Provided that preparation and speed are well aligned, specialty chemicals companies can reasonably expect at least a one-year head start over the local competition.

19.2.2 The Business Model – How to Avoid Common Pitfalls

We have defined four key areas that need to be carefully considered when a company is assessing a relevant business model for China (*Fig. 19.4*).

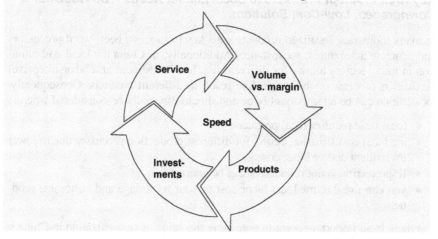

Fig. 19.4. When defining the optimal business model, five areas need to be carefully considered

Key Area 1: Market Share vs. Margin

Generally speaking, the following rule applies. Go for market share only if you are sure that you can beat the local competition on its own turf. This means the foreign company will need to understand or even benchmark local competitors, and to consider how they may react and evolve.

In order to accomplish this, the company has to gain an understanding of how Chinese companies operate. Local competitors mostly operate with "cost plus" pricing. In the local context this means that the price is determined from the cost of goods sold plus the direct production cost plus a suitable margin (around 10 percent in specialty chemicals). Depreciation and overhead costs are usually not reflected in the pricing. As a result, local competitors always, automatically and unwittingly, go for market share.

All facts considered, "going for market share" should only be attempted if either the cost position in manufacturing is not crucial or if, due to scale factors, the cost position of local and international players is the same. A move to specialties is possible if a two- to three-year technology advantage over local competitors can be maintained. In China it would be dangerous if one assumed that a niche would be too small for a competitor to enter.

As in most cases, the local competition comes from the low-quality/low-cost segment of the market, which uses this as a basis to bid for the high end as well. Hence the foreign enterprise will also need to strategically add product lines in China at an early stage in order to defend the low end of the market and to prevent the competition from moving up-market.

Key Area 2: Adapt Products to Local Market Needs – Do Not Offer Downgraded, Low-Cost Solutions

In many industries, localized products are a key to success: beer, word processors and concrete admixtures, for instance. Additionally, in China the local cost situation in many sectors allows for different (i.e. low-investment and labor-intensive) production processes, which therefore result in different products. Consequently, localization can be an important issue and should especially be considered where:

- local taste requires local products;
- the local cost situation allows for different production processes due to lower investment and/or labor costs;
- IP protection is not relevant or can be guaranteed;
- you can use the low local labor cost to your advantage and customize products.

Yet there is an important point to note here: the value of customization in China is often overstated. Chinese customers are rapidly adopting international manufacturing processes and requirements, and are therefore in need of international solutions. Furthermore, Chinese customers, especially in the private sector, exhibit a very high degree of flexibility in adapting their manufacturing processes to raw materials. The claim can be made that "localization" is often only a pretext for selling the same product cheaper. Moreover, if a company enters Chinese markets with localized low-tech products in order to prevent margin erosion or IP loss, we can expect other international competitors to use the introduction of high-end products to China as their entry card, negating the strategy of keeping IP and high-margin products out of China. So the Chinese market needs to be offered the full product portfolio from the start, though the full portfolio does not have to be manufactured locally.

Key Area 3: Investments to Be Based on Long-Term Competitiveness with Locals

International companies often assume that their international business model will also hold true for China. Due to the peculiarities of the local business models already outlined, Chinese companies enjoy advantages which are often underestimated by foreign competitors. Consequently, the competitors' businesses need to be thoroughly understood, and an assessment carefully made of whether and how one can compete with them in a cutthroat environment. In order to explain the inherent advantages of local competitors, in the next section we look at how they arrive at their often surprisingly low prices which are frequently dubbed "irrational", "impossible" or "unsustainable" by their Western competitors.

Excursion: Pricing of Local Competitors, Low but Not Irrational

Even though one can assume that locals usually have 50 percent or less investment costs compared to foreign companies (though larger local companies do not enjoy the same cost advantages as backyard operators do), this is normally not the key to understanding why Chinese competitors can offer such low prices.

Essentially there are two models by which Chinese companies find their prices:

1. Variable cost-based ("cost plus"): In the chemical industry, local players usually consider their raw material costs plus all labor costs (on an annualized, averaged base). Taking these costs as a basis, they add the profit margin they expect (usually 10 percent) and thus reach a sales price. Existing market prices are of little interest to them. If they achieve higher prices, they consider these to be "windfall profits" but nothing to be sustained.
2. In the machinery industry, in the case of simple equipment (switch boxes, for example), the sales price is based on the price of a major raw material, cold-rolled steel for example, and then multiplied by a factor which is in use industry-wide.

While the disadvantages of these systems are obvious, the advantages also merit a closer look. Obviously overhead costs for marketing and technical service can be kept to the bare minimum. Furthermore, depreciation plays no role in pricing. The pricing is based only on cash flow.[5] To neglect depreciation when finding prices actually makes good economic sense in the Chinese context. State-owned enterprises are under no pressure from the financial markets for a return on investment, and therefore the capital basically comes for "free". The private owner usually has no access to bank loans and therefore has to finance the investment with his own capital. As other investment possibilities are very limited for Chinese citizens (no freely transferable currency, no transparent stock exchanges, interest rates on deposits below 3 percent and taxed at the source), even a relatively modest return on capital employed is acceptable.

---

[5] However, for taxation and reporting purposes, Chinese companies naturally consider depreciation similar to their Western counterparts.

Furthermore, the tax rates in China (unusually high for a developing country – 17 percent value added tax (VAT), 30 percent company income tax, around 5 percent other taxes, 10 to 30 percent of payroll cost as social welfare payments) combine with a highly atomized taxation administration to make tax evasion very attractive for many local entrepreneurs.[6] Obviously, international companies cannot resort to large-scale tax evasion, though they usually operate from within special economic zones, which significantly reduces corporate income tax (though not VAT).

How can Western companies compete in this environment? Several possibilities have been used by successful foreign companies in China:

- Scale advantages in production: In many cases, Western companies have entered the Chinese market with factories that are too small, and which in the medium term could thus compete neither with local companies on cost nor internationally on quality due to small volume runs. So world-scale capacity and initial utilization of this capacity for export purposes can deliver major benefits while ensuring the long-term viability of the business.
- Clear value proposition: If the value proposition can be clearly separated from the production cost, either through branding or essential service, Western companies can realize significant advantages over their local competition.
- Lowest possible tax rate: Currently the Chinese taxation system favors the foreign investor, especially if the investment is done in special economic zones, though WTO rules are set to level the playing field in the near future.
- Using financial leverage: Local private competitors need a positive cash flow to survive, especially as they cannot get short-term bank loans. They are thus very vulnerable to delays in bill settlement, and usually cannot buffer variations in raw material costs. This also opens opportunities for market players with superior financing.

Excursion: Investment in Health, Safety, Environmental Protection and Quality Assurance (HSEQ) Is Hugely Important – But Should Be Managed According to Local Requirements

Health, safety, environmental protection and quality assurance (HSEQ) is clearly a key factor for foreign companies in China. For obvious moral reasons, no foreign investor should apply less stringent standards in China than anywhere else. Furthermore, with increasing numbers of international customers starting to produce in China, HSEQ is becoming a differentiating factor to local competitors. HSEQ also represents an investment in the future, since China is tightening its laws relating to workplace safety and environmental protection. And last but not least, law

---

[6] To quote a former editor of the China Business Times, Zhong Dajun: "If every private Chinese enterprise was thoroughly checked, I would say that almost none of them would be able to remain in business. Almost all of them do something illegal, such as tax evasion, bribery or forgery." (*The Guardian* 2003)

enforcement in the area of environmental protection is gradually becoming the norm. Players who have already invested early enough in HSEQ will therefore benefit.

Nevertheless, as HSEQ is a major cost factor, all HSEQ investments in China need to be viewed with a critical eye in order to avoid over-investment. Many HSEQ investments in developed countries are made on the basis of projections regarding expected legislation. The foreign enterprise needs to consider carefully whether such investments are necessary in China. Also HSEQ in other countries has evolved in stages and therefore the existing solutions do not always represent the most efficient approach.

Moreover, where international HSEQ requirements directly contradict Chinese law, it will be the latter that the foreign enterprise should, of course, adhere to.

The HSEQ issues can be summarized thus. Standards for China should not be relaxed, though care needs to be taken that HSEQ costs are kept under tight control, as many local competitors enjoy considerably lower costs in this field.

Key Area 4: Service – An Increasingly Important Factor in Specialty Chemicals

Service is a key element for success in China, and one of the most difficult for Chinese competitors to copy. Due to increasing quality demands, especially for exported goods, as well as increasing quality consciousness among Chinese consumers, local customers are becoming ever more interested in receiving service, and are willing to pay a premium or at least to give significant volume as a reward for service. Service in the context of specialty chemicals usually has to be defined as formulation support. Local R&D is usually not required by the customer.

- Service is the most difficult value proposition for Chinese competitors to copy
- As local customers increasingly try to export, and as local end users become more quality conscious, demand for service increases

- Local R&D not absolutely necessary, but work on the formulation with the local customer is mandatory
- Service for China has to come from China by Chinese
- International image of high standards of service is an advantage and should be maintained

Fig. 19.5. Service is the key to margins and long-term sales in specialty chemicals

Generally speaking, service for China should be performed from a service center in China. This is mainly due to cultural preferences but also because the local market has more stringent speed requirements than other markets. While key service specialists should be localized, there is still a marked preference for expatriate specialists, who are associated with "high-tech". From the latter point we can also conjecture that service has to be in line with the image the company wants to project: modern laboratories, clear procedures etc. We also recommend that the authority level of local service centers be in line with the authority of similar service centers worldwide. Any imbalance in that respect is often perceived as a sleight to the Chinese identity and generally leads to higher personnel fluctuation (*Fig. 19.5*).

19.2.3 Business Relations – Contracts and the "Mandate of Heaven"

To understand business relations in China, we need to understand the partner in his own cultural and economic context. Among European countries, or between the United States and Europe, this context is sufficiently similar and all participants have usually undergone a similar acculturation process (kindergarten, primary school, high school, university), have listened to the same music, had holidays in similar countries and have grown up as part of a meritocratic belief system that is Judeo-Christian in origin. In China, however, the local partner comes from a fundamentally different background. His key managers probably went to kindergarten, where they may have learned the song "Kill the landlord" to the tune of "Frère Jacques". Most high-level and mid-level managers in China would normally have had to interrupt their schooling or academic career to work for several years in the countryside. Often after their return they would have been unable to resume their academic careers. If a Chinese person studied in the 1970s or 1980s, his or her major subject would have been decided by the university, not by the individual. It has to be clearly understood that any Chinese negotiation partner has had a fundamentally different acculturation from his Western counterpart. The dramatic historical changes which especially the generation that is currently in power has seen limits their trust in the future and forces them to maximize current gains rather than future benefits.

Furthermore, Confucianism, though meritocratic, teaches respect for the father and the family above all, followed by respect for learning. This, and the absence of a Judeo-Christian belief system, means that altruism is by and large not a feature of Chinese thinking. Therefore, Francis Fukuyama characterizes the Chinese society as a "low trust society" (*Fukuyama* 1995, p. 75), as trust is mostly reserved for the family. In line with this Confucian model of society the major role of government is not to ensure civil liberties but to ensure balance and harmony. Only by maintaining balance can the government maintain the "mandate of heaven".

The role of trust in Chinese society is illustrated by the graph on the right in *Fig. 19.6*:

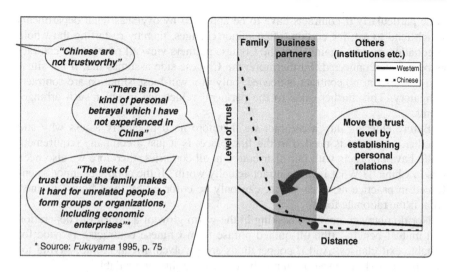

Fig. 19.6. Develop and maintain trust in order to achieve "win-win"

And yet a key feature of business has to be trust, since there has historically been an absence of strong government. Obviously, as mentioned above, the highest trust level is between members of an (extended) family. Nevertheless, business dealings also require trust to be established. As *Reddings* (1993, p. 67) remarks: "Chinese society has come to attach central importance to the notion of trust". Most Western companies, with their rapid rotation of personnel, are ill-suited to establishing long-term, trust-based relationships in business. Nonetheless, trust is a key feature of doing business in China, especially when reaching contractual arrangements. Therefore the key to any business relationship in China is to establish mutual trust. Due to the factors mentioned, this is more difficult then in many other societies, but can also be more rewarding.

What China Really Understands by "Contract"

As Chinese society has different fundamentals than Western society, it comes as no surprise that from a business perspective the biggest issues in China often arise from contracts. In the Western view, a contract (often) precedes relations or can be instrumental in setting up relations or can form the basis for sound relations – and once a contract has been signed, everything is agreed and there is no need, and no room for, further discussion. During the contractual negotiations it is assumed that issues are talked about, and that these can be separated into unrelated parts which can be discussed separately.

The Chinese side, on the other hand, will assume that, if at all, a contract follows relations (and sometimes actually hinders relations). Furthermore, contracts have to be balanced, not only at their inception but also in the future. Consequently, a contract only serves to provide guidelines. As a contract is basically about relations, it is holistic and issues cannot be separated. This can become an issue particularly if contracts have to be approved by overseas legal departments which tend to ask for certain, often minor changes, thereby restarting the whole negotiation process again, since the Chinese partners view all parts of the contract as being interconnected. Furthermore, the Chinese side assumes that if the situation is win-win, no contract is needed; only in a win-loose situation are contracts obligatory. This further adds to the partners' reluctance to enter such arrangements.

Before entering into a contract we therefore need to clearly assess why and whether a contract is needed in the first place. Is it just a company requirement ("We have contracts with our distributors in all countries; therefore we also need this in China")? What is the contract actually worth? If the contract cannot be enforced in practice or, even worse, can only be enforced by the Chinese partner, what is the rationale for it?

Nor do many companies, basking in the warm glow of a "spirit of cooperation and mutual benefit" (the oft-quoted phrase from Chinese contracts), provide for suitable exit clauses. And if cooperations are involved, it can be face-saving for both parties to have a senior, mutually respected arbitrator available.

Detailed minutes of the negotiations are an absolute must. Contractual disagreements frequently arise not from any bad will on the part of the Chinese, but rather as the result of a simple translation mistake. Mutually agreed minutes help. Further, many JVs disintegrate many years after the original negotiators have left the field. Hence, such minutes can prove invaluable for understanding past intentions. A waterproof translation of the minutes and the contract also needs to be ensured. Even high-level joint venture contracts habitually suffer from crude translation errors. Moreover, in China it does not pay to skimp on details, even though the Chinese side may claim that all issues can be resolved amicably after signing.

Furthermore, fairness overrides legality. The Chinese party will often not agree to execute a contract which it considers unfair – and they will not perceive their behavior as unjust.

Another frequently encountered pitfall is that many of the promises made by the Chinese side cannot legally be kept. For example, many economic zones operated by municipalities are technically not allowed to offer tax rebates. Currently, the central government collects company income taxes through the local communities, therefore making rebate schemes on a local level possible, but it is clearly against the rules (*Fig. 19.7*).

Western view		Chinese view
• Contract precedes relation • Once a contract has been signed, everything is agreed • A contract can be discussed issue by issue		• Contract follows relation • Contracts which are unequal are invalid* • A contract provides guidelines* • A contract is holistic, issues cannot be separated

Some practical lessons learned
- Understand why the partner wants to enter the contract, especially in the case of state-owned companies
- If you cannot enforce the contract but your partner can, do not have a contract!
- A co-operation contract has to have a clear exit clause
- For co-operation contracts ensure you have a senior, mutually respected arbitrator
- Fairness overrides legality : The Chinese side will often not agree to execute a contract which it considered unfair
- Check if all promises of the other party are legal (e.g. many economic zones operated by municipalities are technically not allowed to offer tax rebates)
- Ensure a waterproof translation. Ensure that you have a specialist translator during the negotiations (often two, one for legal, one for technical)
- There can not be too many details in a Chinese contract

* This is NOT a legal opinion

Fig. 19.7. The meaning of "contract" in China is different than in the West

Do Not Overestimate the Rationale of Competitors' and Partners' Moves

A classical complaint of Western companies is that Chinese companies do not behave "rationally". As discussed previously, Chinese managers and entrepreneurs have experienced a significantly different acculturation process than their Western counterparts and therefore tend to have a very different outlook on business. In particular, their rationale is not the rationale encountered typically in large, publicly traded companies. Generally speaking, their risk aversion is much lower than in Western companies. In the context of a comparatively unstable situation in the country, which veered widely from feudalism to capitalism to total egalitarianism and to a market economy in the course of just one century, it pays to take risks, since the future is unpredictable even for non-risk takers. On the other hand, where there are risk takers, there must also be losers. The rate of success of Chinese businesses is often overestimated owing to the sheer dimensions of the market and the fact that the less successful players often vanish without a trace. For every local competitor that succeeds there are usually many who did not.

So Western companies should not attempt to learn from their local counterparts about strategy and long-term planning. Rather, they should try to emulate the speed and flexibility of local companies.

19.2.4 Intellectual Property – The Real Killer Topic

China is notorious for its lack of IP protection. As is often argued, cultural roots may partly account for this obvious disregard for the effort of others – copying indicates respect, since copying a teacher is the right thing for a pupil to do. Others point out that China has never had a tradition of commercial invention. We should also remember that failure to acknowledge intellectual property rights is a feature of most developing countries, whether it be Germany in the 19th century or, more recently, Japan and Korea. In many of these instances the nation concerned deliberately pursued a path of "learn and develop", hence in the case of China the strong emphasis of the authorities on "technology transfer".

As long as the legal system in China is not yet fully developed in terms of IP protection, "smart" solutions and careful planning are required. Here are some approaches that may prove useful:

- Production: Can your company bring core components for production in sealed form into the country? This may solve some IP problems relating to high-tech processes. You will, though, need to beware of intrusive customs procedures. A typical example of components brought in sealed are solid-phase industrial catalysts. However, we routinely hear reports that the seals on such catalysts are broken on inspection despite all agreements, and that exhausted catalysts are not, or only partially, returned. If done with sufficient preparation, the foreign company should nowadays be able to protect catalyst know-how, at least for use in its own plants.

- Process: Can you bring in core process steering as "software" without having to expose details such as pressure, flow etc. to prying eyes? In other words, the more the formulation or the process details are brought into the country in computerized form, the less the opportunity for "leakage". There have also been instances reported where key parameters for process control are not kept in China but are steered directly via fixed line from overseas.

- Design: Can you keep the plant drawings confidential? Obviously, the plant drawings are a key IP element. Equally obviously, not everything can be kept confidential as there will be local contractors involved. In extreme cases (new semiconductor fabs, for example), it has been reported that the plant drawings were only available at the site and in computerized form, with each printout of details being individually controlled before it was allowed to be taken off site.

- Raw materials: Can you bring in core raw materials as a "black box"? For chemicals processes especially, it is often possible to bring core parts of the formulation in as a black box. Admittedly, this usually causes its own problems when it comes to adapting quickly to market needs and to changes in local raw material quality. Nevertheless, it has become the method of choice, especially in industries where the formulation is the key know-how.

- Preparation : Do not be unprepared to fight the legal fight. Have good relations to SAIC (State Administration for Industry and Commerce) and to the local labor office in order to get their forceful support if you have to fight for your IP rights or need to enforce non-compete clauses.

- In order to get a full grasp on the issue, IP also has to be discussed and redefined in the Chinese context. In many cases the local competitor comes from a vastly different know-how background. Features which a Western company would rarely deem to be IP-relevant or considers to be already in the public domain are for him highly relevant. A typical example would be HSEQ-related IP, which is becoming increasingly important also for Chinese competitors and is still only loosely protected by Western investors. Therefore, nearly everything should be treated as IP-relevant. Plant and laboratory visits should be limited to an absolute minimum and even simple issues, like the preferred trade journals of the industry, should not be revealed.

And last but not least, it is also good practice to consider where IP can migrate to, and why. IP can either migrate to competitors or via the foreign company's own staff leaving to set up competing plants, especially in industries which are not capital-intensive. This makes it important first to retain key staff and second to keep your physical distance from the competition. Given strong local protectionism, local governments will often support the companies in their area against IP infringements from competitors from other provinces. Furthermore, Chinese employees normally do not like to relocate, as the spouse will also tend to hold a job and the children attend a specific school. Consequently, staff loss to competitors that are not in the same locality is relatively rare. By keeping the number of production sites to a minimum the company can implement further protective measures and improve staff benefits, thereby increasing IP protection.

All in all, there is a whole range of measures available to support IP protection. Only some have been mentioned here. But in summary we can note that IP protection in China demands, above all, constant care when handling IP details, and consistent support from top management.

19.2.5 Human Resources – Finding and Retaining the Right Staff

Finding that All-Important Right Team

China, as a deeply relationship-oriented society, places the highest importance on relationships in all transactions. Therefore, with the right team in China, it is possible to move mountains – with the wrong team frustration and defeat are inevitable. So how do we begin to find the right team in China? In a fast-moving labor market this poses major challenges. The best route to competent staff is via acquisitions. The acquisition target usually has senior staff who are experienced in the industry. Furthermore, employees from state-owned enterprises regard working for a multinational as a clear promotion. And last but not least, these staff have already proven their loyalty. Some acquisitions in China are carried out mainly in order to get key personnel on board.

Another very popular option in China is the public marketplace, mostly the Internet, but also advertisements in newspapers etc. The Internet in particular provides a very cost-effective solution in China today.

Due to the relationship-based nature of Chinese society, suitable candidates can often be located from among personal acquaintances or via other employees. Here a caveat is necessary: this approach often leads to internal entrenched cliques, the hiring of family members and the 'takeover' of the company by groups of closely related people.

The last and frequently used option is to employ the services of a headhunter. A great deal of caution is required here: Many headhunters live by "recycling" their candidates regularly, while others employ clearly immoral methods such as paying a commission to the head of personnel for informing them about potential internal candidates. International headhunters rather than small local agencies are therefore the preferred choice nowadays.

Expatriates

Companies operating in China need to think carefully about whether, and to what extent, they should use expatriates, due to the relatively high costs involved. A clear cost/benefit analysis is therefore a wise move. Usually the rationale for bringing in an expatriate is one of the following:

- To improve local expertise and skill sets: Mostly technical, production or accounting positions, usually for 2 ½ years.
- To ensure a connection to headquarters and to implement company philosophy: Usually general management and senior marketing positions, duration should be at least 5 years.
- To control and supervise: Mostly financial and general management, duration unlimited.

The different objectives listed above need to be clearly analyzed and understood. Especially important are the different continuity requirements of the various expatriate positions.

Many companies try to increase their use of "Asian" expatriates due to perceived cost savings and an assumed cultural affinity with China. This step has to be considered very carefully, as two of the major reasons for having expatriates, namely to ensure a smooth information flow with headquarters and to implement the company philosophy, cannot be expected from Asian expatriates. Furthermore, Asian expatriates in particular tend to face considerably more problems in China than truly international expatriates. This can be attributed to a variety of issues. These include jealousy ("Why does he as a Chinese from Hong Kong get a higher salary than me, a local Chinese, even though we do the same work?"), underestimated cultural differences (how much does a Hong Kong native who speaks Cantonese as his mother tongue and has been to an English school and English university really have in common with someone from Shanghai?) and unacceptable leadership styles (overseas Chinese managers tend to treat their local employees in a more paternalistic and authoritarian manner, while at the same time often efficiently disconnecting any communication between their employees and higher management). Generally speaking, local staff will not accept as many mistakes from people who are fundamentally considered "Chinese" as they will from true foreigners who are thus treated more leniently in this respect. Consequently, despite the obvious advantages of using managers with Chinese cultural roots, their selection requires even greater consideration than that for European or American expatriates.

For many positions, no matter what the particular origins of the expatriates, they should still be a transient feature of business – instrumental in starting the business, but not necessary for the long term, except for key functions. Yet continuity naturally dictates that the expatriates are retained for a certain duration. Current terms (about 2 ½ years in the case of German expatriates) are obviously too short. As a rule of thumb, expatriates require about one year to acquaint themselves with business in China and to build their teams, thus reducing a 2 ½ year stay to 1 ½ effective years.

Strong support from headquarters is absolutely necessary if the aim is for the expatriate to stay in China for longer than 2 ½ years. A clear career path extending beyond the stay in China will need to be established before the posting, in order to justify the long absence from headquarters. Without continuing, non-intrusive support from head office, expatriates cannot provide returns on the money companies spend on them. Furthermore, companies need to take great care not to damage the expatriate in his or her particular function, especially in light of the huge investment the company makes in this person. As previously discussed, Chinese culture places huge emphasis on relations and less on structures. Consequently, measures which are structurally sound in the eyes of an overseas company may turn out to be a relationship disaster. Typical mistakes, for example, include visits by higher management to a local JV prospect without the local general manager being invited to attend as well. This may be appropriate structurally (board level meetings between companies do not necessarily require the presence of local management), and is often accompanied by a benign intention on the part of head office not to overburden the local management. However, it does weaken the local general manager significantly within his organization, as local staff will perceive the arrangement as a clear signal of distrust. Given the near mystical importance of "the big boss" in China, the local general manager will be seen to have lost the "mandate of heaven". Higher management needs to be aware of these idiosyncrasies of local perception and lend full support to local management, especially by giving "face".

To summarize, the high cost of using expatriates has to be well planned and even better executed in order to deliver the required message of management consistency.

Local Staff

Why is staff turnover so high in China? Is this a feature of "the Chinese"? Are "the Chinese" inherently less loyal? The issues are actually not cultural alone but are related to the economy and to management as well. Not only is the labor market in China desperate for certain skill sets, but also, given the still relatively low salary levels in China, a move to a competitor can easily result in a 50 to 100 percent increase in salary. Furthermore, in a country where only 1 percent of the population consists of university graduates, international companies hire from the tiny minority who, in addition to having academic credentials, can also speak English. These people are obviously the best of the best, and have career expectations in line with their qualifications. Very few companies are currently willing to actually offer these opportunities to local staff, and many do not have the local organization to allow for significant career development. Management instability further compounds the issue. With most overseas managers leaving after about 2 ½ years, local staff do not perceive the required commitment from their company, though long-term incentive schemes like health insurance, school fees support, pension schemes, life insurance schemes and similar may help to improve this situation. Moreover, loyalties in China tend to be towards people, not towards companies, a link which is broken far too often through regular management rotations. At the same time, managers are expected to take care of their staff in most walks of life, something which does not fall within the traditional scope of work of Western managers and where they regularly underperform.

Personnel policies usually compound these problems as international companies tend to have well thought-out HR policies based on European or American schemes for compensation and training. Not only are these policies adapted to obviously different cultural needs, they also try to cover a different age structure (usually the age distribution in the Chinese subsidiary deviates significantly from the European or American situation).

It is therefore crucial that the foreign company manages the expectations of local staff very competently, brings a high level of consistency to management and also adapts international HR guidelines to the local situation without compromising the spirit of these policies.

Treat Your Staff Like You Would at Home – But Invest Significantly More Time

The same qualities that make up a good manager of people overseas are also needed to be a good manager of people in China, except that in China these qualities are required in bigger helpings. In China, like everywhere else, leadership capability is the manager's most important asset. Chinese teams can be remarkably creative and hard-working. Unlike Western teams, however, each team member needs clear attention from the team leader. The team leader has also to be very aware of each subordinate's individual personal situation, and will need at times to act as a marriage counselor, lifestyle advisor and much more. Anyone who has ever met high-ranking Chinese officials has a story to tell about their accessibility. This management style has to be reflected in everyday management practice. In China, all relations are personal relations and thus all levels of management will require attention on a personal level. Finally, a manager also needs to be a role model not only in business but in life – in conduct, in everyday life and in terms of honesty.

As discussed previously, Chinese staff yearn for opportunities. Thus a manager will need to fight like a lion for his or her staff, especially with HQ. At the same time, the staff have to be proud of their company. Pride is a hugely motivating factor, especially in countries which so very obviously are status-conscious. If the company is a Fortune 500 company, the Western manager will be aware of this fact in a general way, but the local staff will know the exact ranking of the company.

Quality is often sacrificed for speed in recruitment, especially when setting up a new company, business line or similar. This can be a fatal mistake. Often the best recruits are not those who also speak English the best, nor the ones who understand Western communications, but those who to some extent share Western values. It can be well worth investing the money in an English course for a technically excellent and reliable employee rather than recruiting and promoting on the basis of English skills.

Management consistency is also a huge topic in China. Dishonesty is easily uncovered, and higher management is under much more scrutiny from its own employees in China than anywhere else in the world.

This consistency also applies to personnel policies. These need to be flexible enough to allow for local characteristics, but should not be handled in a way that makes local staff feel like second-rate employees (*Fig. 19.8*).

- Leadership, leadership, leadership
- Build a team – but usually expats do not stay long enough to do so believably and efficiently
- Open opportunities, fight like a lion for your staff, especially with HQ
- Be a role model – in conduct, everyday life, honesty, ...
- Make them proud to work for a global leader
- Be personal, also at board, division or BU level
- Maintain management consistency : Own up to mistakes and expectations
- Introduce long-term incentive schemes

Fig. 19.8. Treat your staff like at home – but invest significantly more time ...

Avoid the Vicious Circle

The most common trap for overseas investors in China results from the vicious circle of overbearing influence from overseas versus overdone localization. Many companies start with a large number of expatriates. Usually, and especially if the China operation is too small in scale, these expatriates become a major cost issue. Additionally, the foreign enterprise perceives a certain moral need for localization: "How can we grow local management talent if we do not give them responsibilities?" As a result, a wave of localizations takes place, leading to premature promotions and management inconsistency. Usually this happens before the company culture is firmly entrenched in the Chinese operation. So, within a short period of time the newly installed local management turns the local organization into a "black box" which is totally non-transparent to overseas management. Key positions are filled with relatives, suppliers are chosen from relatives' companies and customers receive huge price reductions in exchange for significant kickbacks. Profitability, after peaking briefly due to the departure of the expatriates, drops significantly. At some point in time, headquarters then sends in an unsustainably large group of expatriates to clean up, thereby significantly increasing costs.

This vicious circle can only be avoided by good management practice, including a long-term development strategy for key personnel, the use of a matrix organization for Chinese operations as well and constant but non-intrusive supervision. Expatriates should be evaluated on the basis of benefits and not of cost. The same applies to the use of local management talent. As is also clear, companies with a defined and entrenched culture are more immune to such vicious circles (*Fig. 19.9*).

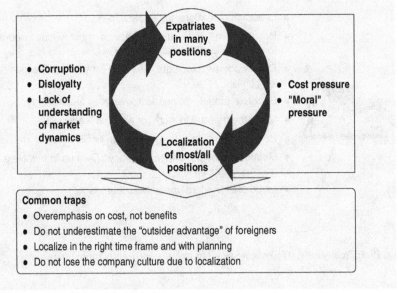

Fig. 19.9. ... and avoid the vicious circle

19.3 The Road Ahead

Though by no means complete, the major concerns throughout the industry, and to some extent beyond, are:

- How to structure the business model (JV/WOFE, holding company...)
- The speed and mode of market entry
- The right business model based on competitiveness with locals, HSEQ, service, product mix
- Relations and relationship management
- IP protection
- Staff and staff retention

We have assembled a list of factors that have proved decisive so far in determining success for operating businesses in the chemical industry in China. They are based on interviews, our own experience and Degussa's extensive experience of manufacturing in China.

There is no master plan for China, nor will there ever be the ten golden rules for China. However, the above are, for most industries, the focal areas that can make or break the business in China. We have highlighted potential solutions – the job now is for each business to evaluate for itself the effectiveness of the solutions we have proposed.

19.4 The Take-Away

China is the greatest business opportunity in the world as we move onward into this century. Deng Xiaoping declared that it is glorious to be rich and now 1.3 billion Chinese want to become rich, and want to become rich fast. The challenge for the chemical industry is to be faster than the emerging competition. So, whatever has to be done in China has to be done not just with speed, but also with a great deal of pragmatism, flexibility and, despite all the speed, patience and respect.

20 DSM in China: In Touch with Evolving Needs in the Specialty Chemicals Market

Stefan Sommer[1]: DSM (China) Ltd., Shanghai, P.R. China

20.1 Introduction

What is required from a global chemical company like DSM in a complex and multi-facet country like China? It is the necessity to deal with a vast amount of change, which in the case of DSM China, has been mostly caused by the far reaching portfolio restructuring of its corporate parent.

Back home in Europe DSM changed from a state-owned company into a public company. It was transformed from a mining operation into a chemical and then a biochemical conglomerate. DSM learned to operate in a fast-paced world, using its strengths in the global market to enter China.

In China DSM has therefore not organically grown to its present size. It obtained many businesses, sites, and employees as a consequence of corporate portfolio changes. Flexibility is a must in a fast-changing economy like China's, and DSM's unique history shows how its core competencies have developed and can be applied again in this new market.

As a consequence, DSM has a flexible investment policy. It likes to give its staff more responsibilities compared to others. DSM expatriates stay longer than average in China and thus demonstrate their commitment to the country. DSM treats their staff more equally than other companies do. Flexibility is the basis of DSM's success in China, and has made it possible to breed a pool of excellent Chinese managers who combine the company's core strategies.

20.2 Royal DSM – From Coal to Biotechnology

From Mining ...

When Dutch State Mines started to operate with only six employees in 1902, it was the start of a key state-owned company in the Netherlands, established to en-

---

[1] The author especially thanks Ari van der Steenhoven, Chief Representative and General Manager of DSM in China from 1992 until 2004. Ari gave valuable input and provided a lot of insights regarding DSM's early growth years in China. Ari started DSM's operations in China in 1992 with only one assistant, and when he retired in 2004 DSM China had grown into an organization with more than 3,000 employees and various joint ventures, wholly owned entities and sales offices.

sure that the country would have enough coal to fuel its economy without having to import from its neighbors. Coal was then, as it is now for China, of vital importance to the country's economic and strategic development, and DSM was poised to be a key player.

From 1906 DSM started to produce coal and mine after mine was opened in the southern part of the Netherlands. In just a few decades, DSM became the most important player in the Dutch coal mining industry, operating next to a small number of private mines. The company also started to develop coal-related chemical products. It was a first diversion into a new industry. In 1927, the American magazine Fortune called the proposal for a fertilizer plant an "obvious Dutch madness". But it was this 'Dutch madness" that was the beginning of a strategy of change and diversification that became the key to the transitions DSM would have to deal with more often in the century to come.

... to Chemicals ...

The focus on chemical products developed slowly as long as the mining industry thrived in the first half of the 20th century. In 1930 DSM produced its first fertilizer, ammonia, at the rate of 120 tons per day. Initially this new business developed at a modest pace because the Second World War and economic hardship cut short development prospects for all of Europe. In 1952 DSM opened its first plant for caprolactam, a product for which it would later gain a leading global position. In the 1950s DSM set up plants for urea fertilizers and the first high-pressure polyethylene plant. Through these investments DSM contributed not only to the economic upswing that began in the Netherlands in the 1960s, but saw also its own business enter a period of high growth. In the early 1960s, the first naphtha cracker started operation and the first melamine plant was built, later to be one of the other core activities of DSM. A second caprolactam plant was opened in Augusta, USA, in 1965.

The ability to deal with change became very urgent in the relatively prosperous 1960s as DSM's mines ran out of exploitable coal and had to close down one after another, the first in 1966 and the last in 1973. In 1969 a full-scale restructuring of DSM took place as it changed into a modern company with six divisions and 33,000 employees. DSM had to earn itself a new place in the economy under entirely new conditions. Its approach to sales changed as, from 1968, it started to license third parties to sell and produce melamine.

... to Biochemicals ...

Another important shift came in the 1970s as DSM expanded into biotechnology. In 1973, the same year as its last mine was closed, DSM started to sell DL-phenylglycine to the Dutch fermentation specialist Gist-Brocades as the key raw material for semi-synthetic penicillins.

Resins followed as products for the coatings industry in 1983. However, the 1980s became the decade of biotechnology for the company with the acquisition of Andeno, a manufacturer of side-chains for antibiotics. Fine chemicals became an additional focus. DSM and Tosoh formed a joint venture, Holland Sweetener, for the production of the artificial sweetener aspartame.

Its listing as a public company in 1989 crowned a period of profound corporate change.

Mergers and acquisitions speeded up in the 1990s and DSM expanded by acquiring the chemical activities of Dutch ACF Chemie and Bristol-Myers Squibb, the polyolefins business of Veba, Vestolen and Dutch fermentation specialist Gist-Brocades in 1998.

... to a New DSM

2000 was another key year in DSM's development as the company called together 100 of its top managers to analyze the company's strengths and weaknesses in a major 'Corporate Strategic Dialogue'. The result of that meeting, called 'Vision 2005', was the blueprint for a new DSM (*Fig. 20.1*).

> Accelerated expansion of the specialties portfolio
> Divestment of DSM Petrochemicals
> Ambitions: > Sales ~ EUR 10 billion
>> Specialties ~ 80% of sales
>> More stable earnings growth
>> Market capitalization expected to double

Fig. 20.1. DSM - Vision 2005

Since then, petrochemicals have been divested and, in 2003, DSM bought the Vitamins and Fine Chemicals division of Roche. This allowed the company to expand its specialties portfolio to 80 percent of total revenue, one of the targets of Vision 2005. The target of EUR 10 billion in revenue for 2005 will most likely not be met and, because of the economic downturn at the beginning of the new century, market capitalization has also suffered. However, DSM is back on the right track. The group posted annual sales of nearly EUR 7.8 billion (including the acquisition of the Vitamins & Fine Chemicals division of Roche – renamed DSM Nutritional Products) in 2004 and generates a strong positive net cash-flow. DSM currently employs approx. 24,000 people in more than 100 countries.

... to a Global Specialties Leader

After a century of successful transformation, the DSM of today is the leading supplier of chemical and biological intermediates as well as nutritional products to the life sciences industries, and has strong market positions in performance materials and industrial chemicals. DSM creates innovative products and services which help to improve the quality of life by being applied in a wide range of end markets and applications such as human and animal nutrition and health, cosmetics, pharmaceuticals, automotive and transport, coatings, construction, housing and electronics.

In 2002 DSM celebrated its centennial and received the title "Royal DSM" from the Dutch monarchy.

DSM today ranks among the global leaders in around 75 percent of its portfolio. It is the global market leader in vitamins, antibiotics, caprolactam, melamine, and many other specialty chemicals and plastics, a key strategy that also explains its current success in those fields in China (*Fig. 20.2*).

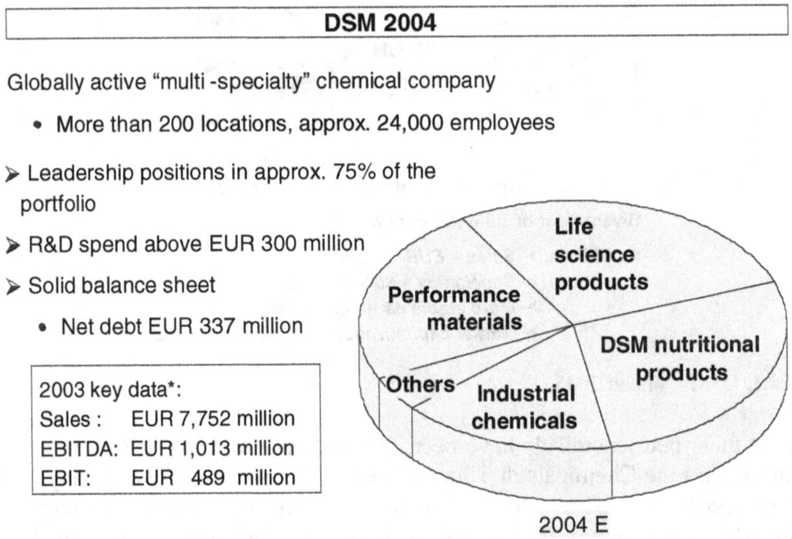
DSM 2004
Globally active "multi -specialty" chemical company
• More than 200 locations, approx. 24,000 employees
➢ Leadership positions in approx. 75% of the portfolio
➢ R&D spend above EUR 300 million
➢ Solid balance sheet
• Net debt EUR 337 million
2003 key data*:
Sales : EUR 7,752 million
EBITDA: EUR 1,013 million
EBIT: EUR 489 million
Life science products
Performance materials
DSM nutritional products
Others
Industrial chemicals
2004 E

Fig. 20.2. DSM 2004

DSM has changed from a state-owned mining operation to the commercially viable biochemical conglomerate it is today (*Fig. 20.3*), from a domestic operator to an internationally renowned company. Many of its current Chinese partners, suppliers and customers face similar monumental changes today, and often aspire to change from a domestic to a global player. DSM has gone through these changes and its expertise and management style is based on its experience in dealing with them. That makes DSM feel at home in China.

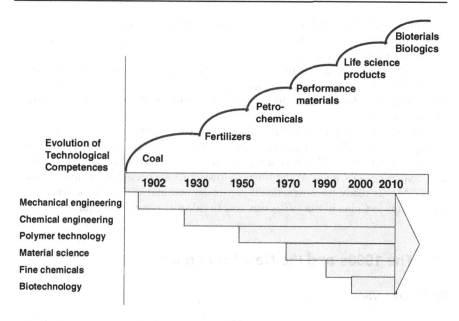

Fig. 20.3. A century of successful transformation

20.3 The Early Years of DSM in China

When DSM in the Netherlands was expanding its range of chemical products and looking for markets abroad, China was an obvious place to go. In 1963 DSM's licensing subsidiary *Stamicarbon* licensed the first grass-roots urea plant with a capacity of 1000 tons per day to China.

Even during the Cultural Revolution (from 1966 to 1976), which made business difficult and isolated China from the outside world, *Stamicarbon* continued to license urea technology to China. In 1978 China's leader Deng Xiaoping said that the country would open its economy again but international business remained cautious. Only when Deng's plans took shape and the first 'Special Economic Zone' was opened in Shenzhen in 1980 did international interest in China re-emerge.

DSM then started to deploy its strategy of licensing local partners for those products for which DSM had a leading position. In 1980, *Stamicarbon* licensed its melamine technology to Sichuan Chemical Works (SCW) for 40 tons per day, the first such deal with a Chinese producer. SCW is today one of the leading melamine producers in China and has constantly increased its capacity over the years.

Sinopec, China's largest oil refinery and petrochemicals conglomerate, became DSM's partner when, in 1986, *Stamicarbon* licensed its leading caprolactam technology to two companies that are now both Sinopec subsidiaries, Sinopec Nanjing and Sinopec Yueyang, for the production of 50,000 tons of caprolactam each per year. That cooperation not only developed into a profitable business deal but was also the foundation of a partnership that has lasted until today.

In 1986, DSM founded its first joint venture in China, an equal partnership with Red Lion from Beijing to set up a factory for powder coatings. Five years later, DSM sold its stake in that company, together with its global powder coatings business, to Akzo.

In 1991, *DSM Stamicarbon* intensified its licensing activities in China by establishing a strategic alliance with one of the country's major design institutes and by deploying consultants. That was at the eve of a major wave of investment by foreign companies in China. At the time, following its successful stock market listing, DSM was ready for international expansion and thanks to its experience and partnerships in China in the 1980s was able to ride this wave successfully.

20.4 The 1990s and the New Millennium

20.4.1 Overview

DSM's four global business clusters: life science products, nutritional products, performance materials and industrial chemicals, are all four represented in China, albeit to different degrees, and make use of the DSM strategy of focusing on achieving global or at least European market leadership for most of its products. China is a cornerstone in DSM's international network that has seen its importance grow year after year since the early 1990s.

A network of representative offices support that work. The first was officially opened in 1993 in Beijing; others followed in Shanghai in 1994, Guangzhou in 1995 and Shenzhen in 1999.

DSM focused initially on increasing and improving its manufacturing facilities in China, often targeting the domestic market. Now it is also becoming an international trader, procuring materials in China from increasingly sophisticated domestic producers to sell them abroad.

But let's go back to the different DSM business groups and their performance in China. To various degrees they all contribute to the current success of DSM. Some – like the resins plant DSM took over from BASF – are even star performers, although they often started as loss-making operations. Thanks to its flexible way of dealing with strategic investments, DSM is now the only fully integrated producer of nylon 6 in China, although not always through majority-owned entities.

DSM's total revenues from China reached more than EUR 420 million in 2004. The company now employs more than 3,000 people in China.

20.4.2 Performance Materials

From a strategic point of view DSM had decided already in the early 1990s, to focus on high-value added intermediate polymers for the growing industries in developed and developing countries such as electronics and automotive. Building on its innovation strengths, DSM has directed its research & development resources to intermediates, and therefore sold industrial consumer products such as engineering plastic forms and shapes, coatings etc. The existing portfolio of the "Performance Materials" cluster consists of engineering plastics, structural resins, coatings resins, as well as elastomers. All these products require detailed knowledge of polymer chemistry as well as broad application know-how to fulfill the requirements of increasingly sophisticated customers.

Performance materials have been a success story within DSM. In two decades it has built up market share through mergers and acquisitions, diversification and globalization. It has created leading positions, also in Asia and China. Except for elastomers which DSM still successfully exports to China, DSM has built a strong presence in all its other performance materials businesses in China.

Its *DSM Engineering Plastics* division was formed in 1992 after the company acquired the engineering plastics business of Akzo in the Netherlands, giving it a position in polyamides and polyesters. The acquisition included compounding operations in Evansville, USA and Stony Creek, Canada.

A foothold in the Asian market was created as early as 1990 with a small team operating from Singapore that had the task of growing the regional market for *Stanyl*, a high-temperature polyamide developed by DSM. At the same time, marketing partnerships were initiated in Japan. DSM's position in Asia was further strengthened by the establishment of new entities in South Korea and Taiwan in 1996.

About midway through the 1990s, the *DSM Polypropylene (DPP)* business group invested in a 50/50 joint venture with Jiangnan Mould Co. in China. *Jiangsu DSM Specialty Compounds (JDSC)* produced a large variety of polypropylene compounds for the automotive, construction and other industries. The location in Jiangyin, Jiangsu province was logical because the molding operation supplying Shanghai Volkswagen was established in the same location. In 1996, the joint venture added polyamide 6-based engineering plastics to its portfolio, allowing DSM to increase its ownership to 60 percent.

Three years later, in 1999, DSM Polypropylenes withdrew from this business and *DSM Engineering Plastics* took over the Jiangyin plant, bought out its partner and changed the name to *DSM Engineering Plastics Jiangsu Co. (DEPJ)*. The site is now the business group's core manufacturing center for its range of polyamides, high-temperature polyamides and polyesters. It serves all the division's needs in China and Southeast Asia.

DSM enhanced its presence in Japan in 1997 through a sales and marketing joint venture with Japan Synthetic Rubber (JSR). In 2003 DSM acquired full control of this company and renamed it *DSM Japan Engineering Plastics*. Earlier, in 1999, *DSM Engineering Plastics India* was established in Pune.

Another signal about the growing importance of the Chinese market in this business group was the transfer of the Asia headquarters of *DSM Engineering Plastics*, responsible for all sites and markets in Asia, from Singapore to Shanghai in 2000. In addition to further strengthening DSM's capabilities to serve customers in China and the rest of Asia, a regional application development center was established in Jiangyin in 2003.

From China *DSM Engineering Plastics* now serves the fast-growing automotive, electrical, electronics and packaging film markets all over Asia. The main products include Akulon (polyamide 6), Stanyl (polyamide 46), Arnitel (copolyester elastomers) and Arnite (thermoplastic polyesters).

DSM Coating Resins (DCR) entered mainland China in 1996 by setting up a 50/50 joint venture - *DSM Eternal Resins (Kunshan) Company Ltd.* - with Eternal Corp from Taiwan as its partner. This illustrates how partnerships can evolve into Asia-wide networks.

DSM started to produce powder coating resins in Taiwan. Eternal was identified as an innovative specialty chemicals producer with a leading position in powder coating resins, state-of-the-art technologies and a strong R&D team. The same company became a minority partner for a manufacturing joint venture in Kaohsiung in Taiwan. Both companies jointly entered China by constructing a facility with an annual capacity of 20,000 tons in Kunshan, Jiangsu province. This plant is presently being expanded significantly. Both the Taiwan and Jiangsu joint ventures are now part of the integrated supply bases of DSM's global powder coating resins business, with additional sites in Augusta, USA; Santa Margarita, Spain; and Schoonebeek, the Netherlands. All sites, including those in Jiangsu and Taiwan, apply DSM's global standards for quality, safety, health and environmental protection and are certified to both ISO 9001 and ISO 14001. Powder coatings will continue to grow rapidly because they are more environment-friendly than conventional solvent-based coating technologies. They even provide special advantages over waterborne coating technologies. China's new emphasis on supporting leading, environmentally friendly technologies will further boost the opportunities for DSM's coating resins business in China.

DSM Composite Resins (DRS) – the largest supplier of unsaturated polyester (UP) in Europe and no.3 globally – acquired the UP resins business of BASF in 1999, and through that acquisition also obtained BASF's Chinese entity Jinling BASF Resins (JBR), a 50/50 joint venture with Jinling Petrochemicals Co, a SINOPEC subsidiary in Nanjing. The joint venture was renamed *Jinling DSM Resins Co. (JDR)* and at the time of the acquisition produced 5,000 tons of structural unsaturated polyester resins and gelcoats each year. However, the market for these products was posting double-digit growth. DSM saw good opportunities and in time succeeded in turning around this loss-making operation. The basic materials are important in the automotive, train components, marine infrastructure and construction industries. *JDR* has increased its annual capacity several times in recent years to nearly 35,000 tons. Further expansions are being planned. In 2003 *DRS* increased its equity ownership in *JDR* to 75 percent.

20.4.3 Industrial Chemicals

The Industrial Chemicals cluster has undoubtedly less strategic importance for DSM than the other 3 clusters. On the other hand DSM is the global technology and market share leader in two key industrial chemicals, caprolactam and melamine; and both products are significant cash contributors to DSM's overall performance. Consequently DSM has concentrated on achieving the same leading positions it had built in Europe and in the Americas, also in Asia; and again China became the key focus area.

For DSM the new millennium in China started with important investments by *DSM Fiber Intermediates (DFI)*. In 2000 DFI acquired a 25-percent equity stake in the nylon 6 plant (annual capacity: 45,000 tons) of Xinhui-Meida, the leading producer of nylon 6 in China that is listed on the Shenzhen stock market.

The new joint venture was named *Meida-DSM Co.* and started operation in 2001. DSM, itself a leading global producer of nylon 6 and the main raw material caprolactam, invested strategically to supply the new joint venture with captive caprolactam, which had previously been imported into China. DSM had been supplying Xinhui Meida since 1993 and the joint venture was a logical way to consolidate this excellent relationship. By taking a minority share, DSM showed again its flexibility in its investment strategy at a time when most foreign companies would only accept a majority share in state-owned enterprises or a wholly foreign-owned venture, if the regulations would allow this.

DFI further expanded in this area by acquiring a 60-percent stake in the caprolactam plant operated by Nanjing Oriental Chemical Corporation of SINOPEC Nanjing. The plant had been using DSM technology since 1986 and is today one of the leading caprolactam producers in China with an annual capacity of 75,000 tons. The company is now called DSM Nanjing Caprolactam Corporation (DNCC).

Thanks to this chain of strategic investment in caprolactam and polyamides, DSM is now the only fully integrated nylon 6 producer in China, although not all entities are majority-owned. Also, not all supplies are captive. DSM is an active merchant supplier, too. However, significant volumes of caprolactam are converted into nylon 6 at Xinhui Meida and some of Xinhui Meida's nylon 6 is used in engineering plastics at *DEPJ* in Jiangyin.

DSM's licensing business, *Stamicarbon*, is still active in China. Its advanced 2000plus™ technology for a grassroots urea plant was licensed in 2001 to CNOOC, China's third largest and most dynamic petrochemicals company. Successfully commissioned in the first quarter of 2004 the plant – located on the island of Hainan with a daily capacity of 2,700 tons – represents the largest single-stream urea plant in China. *DSM Melamine* has announced the start of joint venture negotiations with CNOOC to build a melamine plant with an annual capacity of 120,000 tons at the same location.

In 2002 and 2003, *Stamicarbon* also sold three licenses for major revamps of urea plants to China, each leading to capacity increases of between 40 and 50 percent.

20.4.4 Health and Nutrition

DSM has realized in the early 1990s that especially in the western world, health and nutrition would become key growth areas, fueled by the needs of an ageing and increasingly health-conscious population. In addition to innovative performance materials and cash-contributing industrial chemicals, DSM has therefore engaged in becoming a leading supplier to the health care and nutrition industries, however, with the emphasis on "supplier". DSM never had the resources nor the skills, and therefore never the intention to become a pure pharmaceutical or food company. DSM's core competences in chemistry, chemical engineering and biosciences enabled it, nonetheless, to build its position as the largest global supplier of chemical and biological intermediates to the pharmaceutical, agriculture and food/feed industries. Naturally the global macro trends of health and nutrition are also rapidly emerging in China. Through acquiring the leading antibiotics producer Gist-Brocades in 1998 and the Vitamins & Fine Chemicals division of Roche in 2003 DSM also obtained a strong manufacturing base for these intermediates in China. The "Life Sciences Products" and "Nutrition" clusters have meanwhile also become cornerstones of DSM's China strategy.

20.4.5 Life Sciences Products

Investing in its own manufacturing facilities has marked many of DSM's activities in China. By very carefully picking strong and ambitious Chinese partners, DSM has invested heavily in its future in this challenging market.

However, the key for a manufacturer are the Chinese partners and their qualities. *Shandong Zibo Xinhua Chemferm Industrial Pharmaceutical Co. Ltd* was DSM's first anti-infectives joint venture, created in 1995 in the city of Zibo, Shandong province. The foreign partner, *Chemferm,* was a joint venture between *DSM Anti-Infectives (DAI)* and Dutch fermentation specialist Gist-Brocades and held 51 percent of the joint venture. Shandong Xinhua Pharmaceutical Group, the Chinese partner, held the remaining 49 percent.

The joint venture started to produce cefalexin and cefradine. These are 7-ADCA-based cephalosporins, a family of leading beta-lactame antibiotics. The joint venture was appreciated much by the provincial authorities and won high acclaim. In 1997 it was awarded the title of "foreign-invested enterprise of advanced technology". In 1999, it was awarded the title "best foreign-invested medical enterprise of the Shandong province" and in 2002 "high and new technology enterprise of the Shandong province".

Compared with industrialized countries, China's consumption of antibiotics still lags and the demand for modern products keeps on increasing. In 2000 *Xinhua-Chemferm* increased its capacity significantly and is now one of the leading producers of bulk cephalosporins in China.

The acquisition of Gist-Brocades by DSM in 1998 signaled another turn in DSM's China business. Gist-Brocades, the Dutch fermentation and enzyme specialist, held leading positions in food specialties, antibiotics, yeast, and bakery ingredients. Through this acquisition, *DSM Anti-Infectives (DAI)* became the sole owner of Chemferm, thus gaining a 51-percent share in *Xinhua Chemferm* and a 50-percent share in Zhangjiakou Gist Brocades (ZGB), a joint venture which Gist-Brocades had entered into with Zhangjiakou Pharmaceuticals Co. in 1997. ZGB manufactured penicillin G, the corn- or sugar-based raw material for all beta-lactame antibiotics, as well as 6-APA, a key intermediate, and its derivatives, the leading antibiotics amoxicillin and ampicillin.

ZGB was split into two separate companies in 2001. ZGB remained a 50/50 joint venture between *DAI* and Zhangjiakou Pharmaceutical and concentrates on the production of penicillin G.

In addition to ZGB, Zhangjiakou *DSM Hayao Pharmaceutical Co (DHA)* was formed. Harbin General Pharmaceutical Co., one of the leading Chinese pharmaceuticals producers, was invited to enter the joint venture and took 37 percent of the equity. DSM also holds 37 percent and Zhangjiakou Pharmaceuticals Co. the remaining 26 percent. *DHA* obtains most of its penicillin G from *ZGB* and concentrates on manufacturing the intermediate 6-APA and the semi-synthetic antibiotics amoxicillin and ampicillin.

DHA produces the highest-quality semi-synthetic penicillin on the Chinese market and is even occasionally invited to advise the Chinese State Drug Agency on cGMP-related issues. (cGMP= current Good Manufacturing Practices).

20.4.6 Nutritional Products

Thanks to the acquisition of the global Vitamins and Fine Chemicals division of Roche in 2003, DSM entered the field of nutritional products in China. Also part of the acquisition was the former Roche (China) Ltd. holding company with a number of branch offices, two wholly owned ventures for citric acid and fine chemicals and a majority-owned joint venture for vitamins in China.

The acquired business is now called *DSM Nutritional Products (DNP)* and is run as a separate global DSM division. More than 7,000 employees worldwide generate nearly EUR 2 billion in revenue and the acquisition already contributes significantly to DSM's annual earnings.

Roche Vitamins in China has quite a long history of its own, starting in 1982 when a delegation from the Swiss headquarters visited China. In 1984 the first feed premix plant (Beijing Huadu) with an annual capacity of 200 tons was opened. This was followed in 1987 by Shanghai Xinyang Vet. Concentrates, a feed concentrates joint venture with the Shanghai Agricultural Bureau with an annual capacity of 200 tons. A third feed concentrates plant with the same capacity was later built in Beijing.

The first Roche (representative) office was established in Shanghai in 1993. Two years later, the first joint ventures - Roche Sunve (RSV) and Roche Taishan (RTV) - were established.

In 1996, Roche formed the Roche China Holding in Shanghai. This was very early as holding companies only came into vogue some years later. Roche did so to benefit from the favorable investment and tax incentives which the Shanghai government granted foreign companies setting up holding companies in the city. After acquisition by DSM, the holding was renamed *DSM (China) Limited.*

Two additional joint ventures were created: Roche New Asiatic (RNV) for vitamin B6 in 1996 and Roche Zhongya Wuxi (RZC) for citric acid in 1997. Roche Fine Chemicals Limited (RSFL) was established in 2001 for the innovative green tea extract TEAVIGO.

In 2001, the three Shanghai-based joint ventures (RSV, RTV and RNV) were merged into RVSL - Roche Vitamins (Shanghai) Limited - with two production sites at Gonglu and Xinghuo. The company's main products are VA, VE, and VB6 and it is the first company in China not to use benzene as solvent.

The merger illustrates how corporate restructuring is taking place at DSM. Smaller production units and minority stakes have been changed into majority stakes and consequently merged into a coherent chain of operations.

20.4.7 The DSM China Road Map

DSM has eight joint ventures, three wholly owned companies (*Table 20.1*), and six wholly owned sales offices in China, together employing more than 3,000 people.

Table 20.1. DSM businesses in China

NAME	LOCATION	OWNERSHIP	ACTIVITIES
Xinhua-Chemferm Pharmaceutical Co. Ltd	Zibo (Shandong)	51%	Antibiotics
ZJK GB Pharmaceutical Co. Ltd.	Zhangjiakou (Hebei)	50%	Penicillins
ZJK DSM Hayao Pharmaceutical Co. Ltd.	Zhangjiakou (Hebei)	37%	Penicillins, 6-APA
DSM Engineering Plastics Jiangsu Ltd.	Jiangyin (Jiangsu)	100%	Engineering plastic compounds
DSM Eternal Resins (Kunshan) Co. Ltd.	Kunshan (Jiangsu)	50%	Powder coating resins
Jinling DSM Resins Co. Ltd.	Nanjing (Jiangsu)	75%	Unsaturated polyester resins
DSM Citric Acid (Wuxi) Ltd.	Wuxi (Jiangsu)	100%	Citric acid
Xinhui Meida DSM Nylon Chips Co.	Xinhui (Guangzhou)	25%	Nylon 6 chips
DSM Nanjing Caprolactam Co.	Nanjing (Jiangsu)	60%	Caprolactam
Roche Vitamins (Shanghai) Ltd.	Shanghai	63%	Vitamins, feed/ food premix
DSM Fine Chemicals (Shanghai) Ltd	Shanghai	100%	TEAVIGO green tea extracts

20.5 DSM and China in the Asian Century

The 21st century is already being called the Asian Century. In the year 2002 DSM has developed a China strategy for the years 2003 to 2008. It has been further expanding its "Vision 2005" to include China by strengthening all four strategic clusters, extending its global leadership positions, and exploiting the momentum currently found in China and the rest of Asia. The strategy for China involves three major phases.

The initial years of the strategy (2003 to 2005) are characterized by significant capacity expansions of existing plants in China. Since 2004 (and ongoing) DSM China concentrates on supporting DSM divisions that are still underrepresented and helping them enter the Chinese market. The later years of the strategy will focus on DSM becoming a more active player in the local supply chains in China. The target is at least to double sales by 2008 (from 2002). The key is DSM's human resources strategy, since developing excellent Chinese managers is the basis for any successful business in China.

DSM will continue its successful strategy of expanding in those areas in China where it is already a global market leader, flexibly combining organic growth, acquisitions and alliances.

Since the opening of its first representative office in China in 1993 DSM has grown sales from US$ 20 million to more than EUR 420 million in 2004 including *DSM Nutritional Products,* and has emerged into a leading specialty chemicals manufacturer in China. While still exporting from China, DSM now increasingly also serves the Chinese domestic market with many of its products from its Chinese manufacturing sites, e.g. vitamins, antibiotics, coating and structural resins, engineering plastics, and caprolactam.

DSM is still successfully importing its global leadership products to China from European, American or other Asian manufacturing sites, e.g. melamine and synthetic rubbers. DSM does not see China simply as a low-cost manufacturing location for export purposes, but aims to be a successful, committed, long-term supplier of innovative products from China-based manufacturing sites to the Chinese domestic market. That domestic market still holds enormous, untapped potential for its products.

20.5.1 Expansion of Capacity in China

Of DSM's total business in China (more than EUR 400 million in 2003), 40 percent was accounted for by local production and 60 percent by imports, mainly from Europe. However, sales from local production will increase rapidly in the next few years. This will be achieved through a number of strategic projects which DSM is presently undertaking in China.

DSM Fiber Intermediates is expanding its caprolactam capacity at DSM Nanjing Chemical Company (Jiangsu province) from 75,000 tons per year in 2003 to

140,000 tons by 2005. Construction for this project has already started and will allow DSM to further grow its already strong market share in China.

Since 1999, *DSM Engineering Plastics* has twice increased its compounding capacities at its Jiangyin plant in Jiangsu province. A state-of-the-art application development facility has been opened at the same site. This will allow DSM to accelerate its market penetration as a supplier of high value-added engineering plastics to fast-growing markets in China such as the automotive, electric and electronics, building materials and packaging films sectors.

DSM Composite Resins has announced an increase in capacity of unsaturated polyester resins in Nanjing by at least 50 percent within the next three years.

DSM Melamine has announced that is has entered into joint venture negotiations with CNOOC, the third largest Chinese petrochemical company, to build a global-scale melamine plant with an annual capacity of 120,000 tons near the CNOOC urea plant on Hainan Island. *DSM Stamicarbon* licensed this urea plant in 2001.

Late in 2004 DSM announced a major strategic alliance with one of the leading Chinese pharmaceuticals companies, North China Pharmaceutical Group Company (NCPC). DSM will buy a minority share of NCPC Ltd. which is NCPC's core affiliate listed at the Shanghai stock exchange, and will establish global scale joint ventures with NCPC in the area of vitamins and antibiotics.

20.5.2 Developing New Markets in China

Compared with their global market shares, some of DSM's businesses are still underrepresented in China. The company is now working on market strategies to improve the positions of these businesses in China. They include *DSM Fine Chemicals, DSM Pharmaceutical Products, DSM Dyneema* and *DSM Food Specialties.*

DSM has become an integral part of the Chinese industrial manufacturing and pharmaceutical supply chain, based on its broad knowledge of the Chinese market and long-term customer relationships. This gives DSM the possibility of exploiting the business opportunities offered by the Chinese market, such as the ever growing demand for transport, feeds, functional food, dietary supplements, drugs, personal care products, medicines, communication systems and housing. DSM sees vast growth potential in all those sectors to which the chemical and pharmaceutical industries are major suppliers.

The Asian Century has begun and DSM will use its talented people as well as its leading products and technologies to participate in the unlimited opportunities.

20.6 China and DSM – Managing the Future

In the past decade the chemical industry in China has seen double-digit growth every year, much higher than the official annual growth of GDP. As a witness to this unprecedented growth, DSM is eager to participate in this very promising sector. DSM expects sustainable growth for the near future, but this is unlikely to be linear. For that reason, apart from being first and foremost an enthusiastic participant in the Chinese market, DSM will also carefully manage its investment policy and the related investment risks. Laws and policies set by central government show a keen awareness of the dangers threatening the country's development, but often the leverage to push through necessary changes in a timely fashion is lacking. What possible problems do we see?

20.6.1 In Health Care

The emergence of the SARS virus in 2003 served as a much needed wake-up call. While both the humanitarian and economic damage caused by the SARS epidemic were limited compared with earlier assessments, it showed that there are other very realistic risks around the corner.

AIDS/HIV might be just another health danger that fortunately has been recognized as such by the current government. The stringent approach adopted by the government at the beginning of 2004 has been much applauded, but we still face a potential time-bomb that is waiting to be defused.

On a general level, the lack of health care and knowledge in rural areas still provides a dangerous climate for these and other diseases that can explode into disruptive epidemics.

20.6.2 In Energy and the Environment

Current economic growth, combined with rather inefficient energy consumption and rising oil and gas prices, might have an adverse influence on our industry as prices are rising constantly. The brownouts in the summers of 2003 and 2004 were an early warning of a risk that has been too long ignored.

For oil, China is dependent on the world market, but it has huge reserves of coal. Coal gasification is inexpensive and will provide the basis for a new wave of industrial development. Several plants have started up but suffer greatly from the current lack in railway capacity.

Great progress has been made in controlling pollution and in finding more efficient ways of using natural resources in the main cities, but just as in the case of many other serious threats to the country's stability, only the first important steps have been made. Polluting industries have been moving to smaller towns, where it is harder to enforce environmental regulations.

20.6.3 In Human Resources

The sudden shortage of migrant workers in the Pearl River Delta in 2004 was a clear illustration of how closely rural poverty and economic growth are interlinked. Central government's successful efforts to deal with the income gap between urban and rural areas have been one of the reasons for migrant workers to stay at home. In a huge and highly complex country like China, even positive measures can trigger side-effects that may have a temporary negative influence on other parts of the economy.

The labor shortage reflects how the widely discussed income gap between urban and rural areas and migration can not only cause social unrest, but also large-scale shifts in the labor markets that are hard to control.

The economic diversity of China makes it almost impossible to give an assessment that is valid for the whole country. Labor costs in Shanghai are about five times higher than in other parts of China. Some national figures for inflation, GDP etc. are of limited value. Wages on the east coast are not only increasing in general. Huge investments by BASF, Bayer, Shell, Exxon and BP, which each plan to invest EUR 3 billion or more in China in the coming years, add to the pressure on that section of the labor market on which we also rely. Many people will find employment, but large amounts of engineering capacity will be absorbed, too. The shortage of labor might therefore become a problem in the more educated labor force from which DSM and others recruit.

Significantly lower labor costs, about one fifth of those in the West, make China an attractive place to invest, although productivity is also lower. In addition, plants can be built at considerably lower cost in China, depending on a company's ability to deal with local sourcing and local contractors.

20.6.4 In the Market Economy

Corruption remains an important problem facing the government in its efforts to move the country to a market economy with a more efficient allocation of resources. High-profile corruption cases in the media show concrete attempts by the government to deal with the problem. However, more such efforts must trickle through to day-to-day economic activity.

Chinese banks are also having trouble moving away from their past as instruments of the planned economy. Their non-performing loans – after a few years of reduction – seem to be on the way up again, increasing the vulnerability of the financial system. The government has been offering guarantees to safeguard the private citizen's deposits, thus avoiding a major collapse. Large-scale panic runs on banks have been avoided, but seem possible. The current status of the financial system does not seem sustainable in the long term, as China moves closer to a true market economy. But also in this sector we see fundamental change, as foreign banks actively buy into Chinese banks and apply for wider business licenses.

The lack of protection of intellectual property is another indication that the road to a real market economy is a rocky one. Again, on a central level, efforts in setting laws and policies are rather impressive. However, on a working level, the loss of intellectual property as a result of staff departures is still a problem.

China has improved the protection of intellectual property since joining the World Trade Organization in 2001, with the Chinese Patent Management Office, the Industry and Commerce Administration and the Trademark Office working together to enforce IP legislation. DSM welcomes and supports such moves, since they mitigate the risks of investments and will allow multinational companies to bring more and more advanced technologies to China.

20.7 Conclusion

Doing business in any part of the world involves risk and DSM does not see China as generally presenting a higher risk than other countries. The bigger threat would be if DSM did not seize opportunities, internationally or locally, while the competition did.

None of the risks are likely to escalate suddenly; they are here now but are contained and managed reasonably well by central government.

Balancing these risks against the huge potential offered by the Chinese market is a challenging task for all stakeholders interested in China, whether these are multinational companies, managers, politicians, scholars or scientists.

DSM is aware of the risks and includes thorough risk assessments in its growth plans for China. The company is already well-positioned in China and has ambitious plans. Executing these plans with the right amount of caution will further solidify DSM's position as a leading international specialties company in China.

21 Vitamins – Opportunities and Challenges for Both Western and Chinese Producers

Manfred Eggersdorfer: DSM Nutritional Products Ltd., Basel, Switzerland

21.1 Vitamins Are a Dynamic Market

Vitamins are essential for growth and health of humans and animals. As they cannot be synthesized in the body, vitamins have to be provided via diet and nutritional supplements. The market for vitamins developed to a value of US$ 5 billion over the last seven decades. The major application is in animal nutrition and health with a share of 40 percent, followed by human applications with a share of 39 percent. Smaller amounts are used in the personal care industry and in a number of technical applications with a distribution of 5 and 6 percent respectively (*Fig. 21.1*).

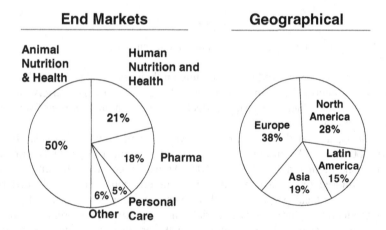

Fig. 21.1. The vitamins market including information about sales according to regions

In recent years the various segments recorded an average annual growth rate of three to four percent. The market segments can be further divided into sub-segments, each having its own dynamics. Geographically, the highest vitamin consumption is in Europe, followed by North America, Latin America and Asia.

While the market has shown volume growth for decades, from a manufacturing viewpoint there have been ups and downs. The dynamics in this business are determined by innovations in processes and applications and, in the last two decades, by the increasing role of Chinese manufacturers.

21.2 Vitamins Require a Complex Production Set-Up

Vitamins are produced on an industrial scale by chemical synthesis or partial synthesis, by fermentation or by extraction from natural materials. Chemical synthesis is still the dominant method. However, more and more fermentation processes are being implemented (*Fig. 21.2*).

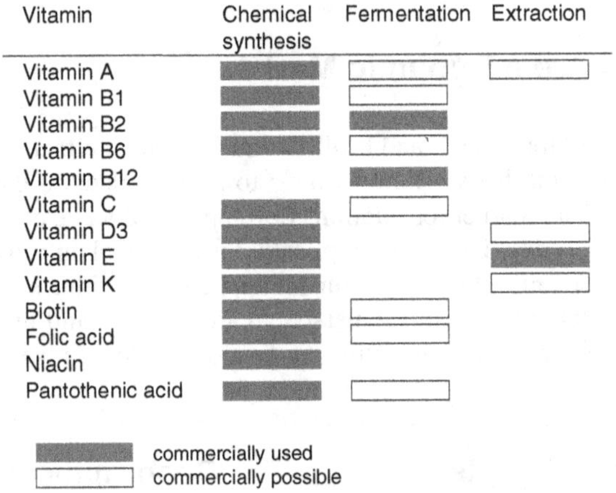

Vitamin	Chemical synthesis	Fermentation	Extraction
Vitamin A	▬	☐	☐
Vitamin B1	▬	☐	
Vitamin B2	▬	▬	
Vitamin B6	▬	☐	
Vitamin B12		▬	
Vitamin C	▬	☐	
Vitamin D3	▬		☐
Vitamin E	▬		▬
Vitamin K	▬		☐
Biotin	▬	☐	
Folic acid	▬	☐	
Niacin	▬		
Pantothenic acid	▬	☐	

▬ commercially used
☐ commercially possible

Fig. 21.2. Different technologies compete in the production of vitamins

Vitamins are complex products depending on their chemical structure. They are divided into water-soluble and fat-soluble classes. Due to the variety in their chemical structures, their industrial production requires a broad array of scientific and manufacturing competencies.

In addition, annual demand worldwide varies from volumes between 10,000 and 100,000 tons for Vitamin C and Vitamin E to a number of products with consumer demand in the range of 1,000 to 10,000 tons. Others are in the range of only a few to hundred tons. Each has its own synthetic route, completely different physical and chemical properties, and its own requirements in terms of aspects such as production volumes, raw materials and handling. Therefore, a number of special technologies have been developed for the production of vitamins. In addition, a number of vitamins are sensitive to heat, water, light and other factors. Special arrangements therefore have to be made to stabilize the final product utilizing so-called product forms.

21.3 New Entrants Result in a Competitive Environment

From 1933 when Roche achieved the first industrial production of a vitamin using the Reichstein synthesis for Vitamin C, industrial production processes and their applications have been developed. Since the early1960s, all vitamins have been

accessible by synthetic methods. With the growing economic importance of vitamins, more and more companies entered the business. BASF, Merck, Duphar,
Rhône-Poulenc, Lonza and others developed their own synthetic approaches for
manufacturing vitamins. Furthermore, companies with fermentation expertise used
their competence to enter the market at relatively low investments. However, innovative companies continuously improved their production processes and used
this to increase their market share

Beginning in 1980s, the first Chinese companies started up small vitamin production plants (*Fig. 21.3*). They built on synthetic routes available in the public
domain and cooperated with Western brokers to market the products, preferably in
the West. They initially marketed the products in the animal nutrition and health
segment where the quality and regulatory requirements were lower. Due to cheap
labor costs, low investments and other low costs like depreciation compared to
traditional producers, they were successful in establishing their own independent
positions.

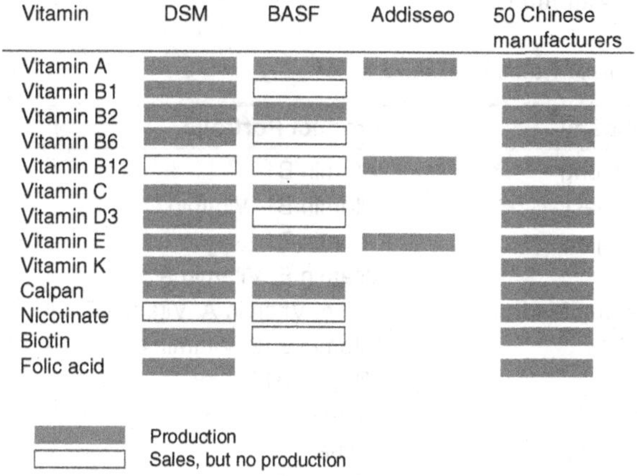

Vitamin	DSM	BASF	Addisseo	50 Chinese manufacturers
Vitamin A	Production	Production	Production	Production
Vitamin B1	Production	Sales, but no production		Production
Vitamin B2	Production	Production		Production
Vitamin B6	Production	Sales, but no production		Production
Vitamin B12	Sales, but no production	Sales, but no production	Production	Production
Vitamin C	Production	Production		Production
Vitamin D3	Production	Sales, but no production		Production
Vitamin E	Production	Production	Production	Production
Vitamin K	Production			Production
Calpan	Production	Production		Production
Nicotinate	Sales, but no production	Sales, but no production		Production
Biotin	Production	Sales, but no production		Production
Folic acid	Production			Production

Production
Sales, but no production

Fig. 21.3. The competitive environment in the field of vitamins

The key vitamin producers Roche Vitamins (acquired by Dutch company DSM
in 2003) and BASF evaluated the Chinese competitive situation early and entered
the Chinese market by establishing joint ventures and building up their sales organizations. Both companies engaged with investments in new plants together
with their respective partners. However, the creation of additional capacities by
new Chinese entrants resulted in overcapacities for all vitamins. The resulting
price pressure initiated a consolidation of the industry. A number of Western producers stopped vitamin production and also smaller Chinese manufactures could
not build up a sustainable position. The key players reacted by investing in world-
scale plants. Overall, this development had a drastic impact on the market structure.

As summarized in *Fig. 21.3*, today only DSM is a full producer/supplier of all
vitamins to the above-mentioned industry segments. BASF´s strategy is to focus
on vitamins which offer a competitive position by technology leadership based on

the so-called Verbund (network of raw materials and intermediates for Vitamin A and Vitamin E) and synergies in fermentation (Vitamin B2). Only a few additional Western producers remain with a position in only one or two vitamins focusing on niche segments and/or partners.

The Chinese manufacturers have evolved a smart strategy. Most of the production processes have been developed by research institutes and universities like the Shanghai Institute for Pharmaceutical Chemistry or Fudan University. In the past, when the companies were state-owned and there was no local competition, such technologies and processes were available to all potential producers. Especially pharmaceutical companies became involved in the field and established production facilities. Nowadays, Chinese producers are established for all vitamins. At the early stage, the Chinese manufacturers were using low-tech approaches with operations far removed from optimum use of raw materials and energies. However, development is rapid and Chinese producers increasingly overcame technical limitations. Thus, nowadays, they are able to compete in the manufacture of complex products like Vitamin E.

Table 21.1. Key Chinese vitamins producers

Company	Product Portfolio
Huazhong	Vitamin B1
Northeast	Vitamin B1, Vitamin C
Zhongjing	Vitamin B1
NHU	Vitamin E, Vitamin A
Zhejiang Medicine Co	Biotin, Vitamin A, Vitamin E
NCPC	Vitamin B12, Vitamin C
Hubei	Vitamin B12
Xinfu	Calpan
Lion King	Calpan
Desano	Vitamin B2, Biotin
Guangii	Vitamin B2

Table 21.1 details the major vitamin manufacturers. Overall, there are over 50 indigenous companies operating. One of the biggest is Zhejiang Medicine Co. Due to low labor costs as well as low investments in small and simple plants; they are competitive even with lower yields and smaller campaigns. *Table 21.2* gives assumptions of the potential cost structure. Raw materials are less expensive in some cases. However, this advantage is partly negated by low yields. Nevertheless, raw materials are a major cost driver and Chinese producers will improve consumption and operations to offset raw material price increases.

Primary energy costs are about the same as in the West but due to a lower degree of optimization, energy usage is usually higher in Chinese factories. Waste disposal costs, labor costs and fixed costs like depreciation tend to be lower compared to Western producers. This difference may be significant and often result in a cost advantage even if the Chinese producers apply inferior technologies. Global changes like the increase in the price of oil and other raw materials have a greater impact on their cost position and are the main cost driver. Chinese producers have now started to make substantial efforts to improve their technologies and offset raw material price increases.

Table 21.2. Assumed cost structure in vitamins production

Raw materials	50-70%
Energy costs	12-16%
Waste costs	1-3%
Labor costs	4-8%
Other costs	10-15%

21.4 New Breakthrough Technologies Impact Competitive Position

The key players have developed specific technologies for the production of vitamins. However, in most cases existing routes are no longer protected by basic patents. The long sequence of chemical transformation results especially in vitamins with low volumes (10 to 1,000 tons per annum) in high investment and unit labor costs. The common denominator in all approaches is innovation as the main driver to reduce production costs by fewer steps and higher yields, resulting in lower raw material and energy consumption. Chinese producers started to enter the market by implementing established synthetic routes. With more experience and the drive of universities to operate in the international scientific league, new technologies are also being developed in China.

One example of how the development of breakthrough technologies impacted the competitive scenario is Vitamin C. *Fig. 21.4* provides an overview of the different technologies. The currently favored Vitamin C technology is known as the Reichstein process (shown at the top of *Fig. 21.4*). This uses a combination of fermentation and a number of chemical steps. It has been optimized over the years to exploit its full potential.

Biotechnology has the potential to shorten this process chain by extending the scope of transformation using microorganisms. The first achievement was the ketogulonic acid process (shown in the middle of *Fig. 21.4*), originally discovered by DSM and later independently by a Chinese university. This breakthrough technology combines several steps in the conversion by two microorganisms and a chemical transformation. It was first implemented by Chinese Vitamin C producers and gave them the capability to become cost-competitive, even in smaller plants with capacities of 5,000 tons per annum, compared to the established Vitamin C producers like DSM, Takeda or Merck which operate plants with two to six times the capacity. With this advantage in place, the Chinese Vitamin C producers ramped up the market and stimulated further consolidation. As a consequence, most of the Western producers which had no access to the new technology left the market. However, a number of smaller Chinese producers also had to close down and only four of them are currently left in the market.

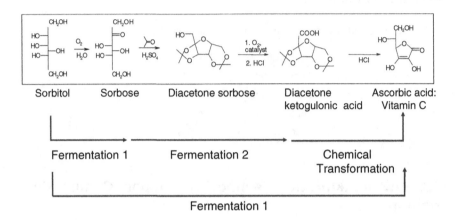

Fig. 21.4. Vitamin C production technologies

This situation energized DSM as world leader in vitamins (including Vitamin C) to evaluate even more elegant and cost-competitive routes. Recently a breakthrough was announced by programming a special microorganism which allows full conversion from the raw material to Vitamin C in only one step (shown at the bottom of *Fig. 21.4*). The impact is tremendous. While the ketogulonic acid process requires two fermentation steps and a subsequent chemical conversion, direct fermentation needs only one set of fermenters, separators, centrifuges and dryers. This reduces both the necessary investment and complexity of the plant including operations and will result in a new benchmark for cost leadership in Vitamin C.

21.5 Product Forms Are an Additional Differentiator

Other differentiators for vitamins are product forms and new applications. Form technologies are essential tools for stabilizing vitamins, protecting them from oxygen, moisture or heat, making them bioavailable or generating product forms for different applications ranging from tablets to beverages.

Fig. 21.5 exemplifies the role of product form technologies. While the principles for the chemical synthesis or fermentation process for vitamins are in the public domain, form competences are often secret and the qualifier for differentiation from competitors and cooperations with customers.

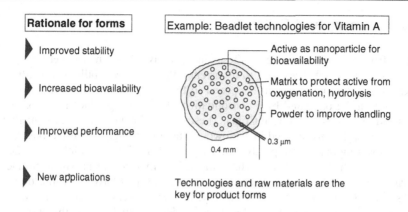

Fig. 21.5. A number of reasons for product forms

The potential of product forms achieved by applying different technologies or matrix materials is demonstrated in *Fig. 21.6*. Vitamin C can be used in different applications resulting from tailor-made forms like:

- Ultra fine powders for antioxidants in food
- Granulate in tablets
- Ascorbyl palmitate as an emulsifier
- Vitamin C polyphoshate (Stay C 35) for aquaculture
- Vitamin C monophosphate (Stay C 50) for cosmetics.

Each form has a unique property profile tailored for the intended application. This adds value for the customers and is a barrier for the competitor because he has to match the standard. For Chinese manufacturers this is a hurdle. They still sell much of their production via brokers and miss direct access to customers. The strategy to copying product forms from market leaders works only in part because a number of technologies are still patented and the leaders regularly improve their forms.

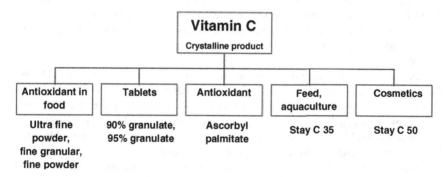

Fig. 21.6. Product form technology can deliver tailored forms to meet the needs of different industries

21.6 Summary and Outlook

Starting from the first lab-scale syntheses in the early 1930s, the market for vitamins was created and developed to a current value of US$ 5 billion (ingredient market size). The major growth happened in the last three decades and was accompanied on the one hand by a consolidation of Western producers and on the other hand by new entrants from China. These concomitant developments (market growth, consolidation) resulted in a growing, but still strongly contested business for the remaining players. In parallel, major progress in new technologies resulted in lower production costs. Due to the competitive environment in which production capacities for all vitamins exceed consumption, prices have been in a downward spiral for several years. Consequently, further market consolidation is to be expected. Survival is assured only for those producers which follow the optimum strategic mix of developing advanced production processes and technologies, using the most advantageous production location (not in every case will this be in China) and entering strategic alliances wherever this is appropriate and will improve the competitive situation of the respective partners.

Chinese manufactures now offer the complete line of vitamins. Currently, they still lag behind their Western competitors in process chemistry and technology. However, they are quickly closing the gap. There are over 50 indigenous companies producing vitamins in China. From their first attempts to start vitamins production on the basis of publicly available know-how and expired patents in usually low-tech production facilities about 20 years ago, they are now starting to compete in complex vitamin production processes like Vitamin A and Vitamin E.

The success story of Chinese manufacturers (growing quickly, winning market share with low prices) is based on a number of specific advantages including low labor costs, relatively low investment costs, easily accessible capital, in some cases lower raw material costs and the ability to offer their goods at close to production cost. However, it can be observed that some of these special Chinese effects are starting to change, leading to a consolidation among the Chinese manufacturers themselves, more restricted capital use and more rational pricing.

To remain successful in this kind of business, other requirements must be fulfilled beyond just offering products at the lowest conceivable price. Advanced technology, optimized production scenarios, the right product portfolio, innovative strength and a smart understanding of the market are some key elements of a long-term strategy.

Vitamins are still a very dynamic business with a lot of new science in chemical synthesis, fermentation, applications and health beneficial effects. Increasingly, science is pointing to the role of vitamins in nutritional networks with other bioactives. This provides innovative companies with new fields of applications for vitamins and health beneficial compounds as well as creating new patentable intellectual property and thus fostering the competitive edge.

After more than seven decades the vitamin business remains an exciting field offering opportunities and challenges in science, technology, production and marketing for both Western and Chinese manufacturers.

Authors

Eric Baden holds an MBA degree from Warwick University in the United Kingdom. He joined Degussa AG in April 2001 as President, China Region. Over a period of 15 years Mr. Baden has accumulated profound regional expertise on various assignments across Asia, particularly in the areas of joint venture negotiation and change management. On his most recent assignments prior to joining Degussa AG, Mr. Baden was general manager of Thermphos Xuzhou Co., Ltd. a Sino-Dutch joint venture for phosphoric acid and phosphorus salts, and headed New Business Development at Hoechst South Asia Ltd. in Mumbai, India. He is an advisor to the Qingdao Economic Trade Development Zone and to the Jiangsu provincial government. He is also a member of the executive committee of the Association of International Chemical Manufacturers (AICM) and of the advisory council of the European Chamber of Commerce in China (EUCCC), as well as Chairman of the Intellectual Property Work Group of the Executive Committee of Foreign Investment Companies (ECFIC).

Degussa (China) Co., Ltd., 16/F, Beijing Sunflower Tower,
37 Maizidian Street, Chaoyang District, Beijing 100026, P.R. China;
Phone: +86 10 8527 6401, Fax: +86 10 8527 5987,
E-Mail: eric.baden@degussa.com

Michael Brueckner is a partner in the Global Health & Life Science Practice of Accenture in Frankfurt, Germany. He is a member of Accenture's Strategy and Business Architecture Service Line. Mr. Brueckner has broad international experience within the area of prescription, generic and OTC drugs with a focus on corporate and business unit strategy development and implementation, capability development and organizational restructuring covering all main functions (research and development, supply chain, sales and marketing, and finance and controlling). Mr. Brueckner joined Accenture in 1999 after working for more than four years in the Healthcare practice of Arthur D. Little across Europe, the United States and Asia. Additionally, he was a member of the ADL Balanced Scorecard and Customer Management Competence Groups.

Mr. Brueckner studied business administration at the European Business School (Germany), UC Berkeley (USA), ESC Dijon (France) and at the Kellogg Graduate School of Management (USA), where he obtained an MBA degree. He has contributed to a large number of industry seminars and published a series of articles and white papers concerning, for example customer profitability management, challenges and strategic options in the Indian pharmaceutical market, CRO management in the Internet era and optimization of European distribution networks.

Accenture GmbH, Campus Kronberg 1, 61476 Kronberg, Germany;
Phone: +49 6173 94 67245, Fax: +49 6173 94 47245,
E-Mail: michael.brueckner@accenture.com

Dr. Xiangdong Chen received his master degree in economics in 1988 and his bachelor degree in engineering in 1982 at Northeast University in China. He is currently a full professor in international technology transfer and international finance at the School of Economics & Management, Beijing University of Aeronautics & Astronautics. He has many years of research and working experience on innovation and international technology transfer – including a practical survey covering 53 British companies transferring technology to China when he was a visiting scholar at Manchester Business School; German companies' technology activities in China when he was a visiting professor at the University of Applied Science in Brandenburg, Germany; the technology resources positioning of Japanese companies in China and linkages in technology resources among two sides and four localities when he was a visiting research fellow at City University, Hong Kong. He currently heads the Department of International Business Management at the School of Economics & Management, Beijing University of Aeronautics & Astronautics. His major research interests cover fields such as multinational companies in China, international technology transfer and regional innovation studies. He has published a large number of academic papers in international journals and international conferences. In recent years, his innovation research interest has extended to the pharmaceutical sector and includes the practical survey and empirical study of emergence innovation in the pharmaceutical industries during the SARS epidemic in 2003, and the patent-based technology positioning of companies from typical countries and districts.

The School of Economics & Management, Beijing University of Aeronautics & Astronautics, Beijing 100083, P.R. China;
Phone/Fax: +86 10 8231 5597, Fax: +86 10 8231 6100,
E-Mail: chenxdng@buaa.edu.cn

Dr. Manfred Eggersdorfer is global R&D director for DSM Nutritional Products, the world leader in vitamins, carotenoids and nutritional ingredients for the human and animal nutrition industry. Prior to this he worked for BASF in various positions including head of Research and Development Fine Chemicals. Manfred Eggersdorfer studied chemistry at the Technical University of Munich and did post-doctoral studies at Stanford University, California. He is a member of the Curatorium of the Fraunhofer Society and author of numerous publications in the fields of vitamins, innovation in nutritional ingredients and renewable resources.

DSM Nutritional Products Ltd., Research & Development, P.O. Box 3255,
4002 Basel, Switzerland;
Phone: +41 61 687 19 53, Fax: +41 61 688 50 99,
Mobile: +41 79 244 90 06, E-Mail: manfred.eggersdorfer@dsm.com

Dr. Nicole Feling joined Bicoll as public relations manager in September 2001 and is responsible for all media and public relations. Dr. Feling has an extensive scientific background but specialized in the field of public relations. She has excellent knowledge of the press market and, prior to joining Bicoll, gained special insights into the U.S. press market during 1½ years working as a freelancing journalist in San Diego, specializing in reports about medicinal and pharmaceutical topics. She received her master degree and PhD in chemistry from the University of Munich. Her work was closely related to the isolation, identification and elucidation of medically relevant natural active agents.

Bicoll Biotechnology (Shanghai) Co. Ltd., Bibo Road 518A,
Zhangjiang Hi-Tech Park, 201203 Shanghai, P.R. China;
Phone/Fax: +86 21 3895 3452, E-Mail: nicole.feling@bicoll-group.com

Dr. Gunter Festel founded the consulting and investment firm Festel Capital in early 2003. His company specializes in supporting spin-offs and buyouts in the chemical, pharmaceutical and biotech industry from the first conceptual steps through transaction support to business optimization and the execution of exits. Prior to this, he was head of the Chemicals and Healthcare Practice and a member of the Management Committee of Arthur D. Little in Switzerland. His first years in consultancy were spent with McKinsey & Company in Brussels, Frankfurt and London. Before working as a consultant he was employed at Bayer AG as head of an R&D laboratory, assistant to a member of Bayer's Management Board and product manager with responsibility for a EUR 150 million business.

After his military service Gunter Festel studied chemistry with a focus on biochemistry and polymer chemistry (master degree and PhD), as well as business administration and economics (master degree). He is chairman of the Association for Chemistry & Economics within the German Chemical Society (GDCh) and is the author of numerous books and publications on the subjects of chemical economics, M&A and buyouts as well as innovation management, biotechnology and genetic engineering.

Festel Capital, Schuermattstrasse 1, 6331 Huenenberg, Switzerland;
Phone/Fax: +41 41 780 1643, Mobile: +41 796 527 112,
E-Mail: gunter.festel@festel.de

Dr. William A. Fischer is a professor at IMD (International Institute for Management Development) in Lausanne, Switzerland, where he specializes in issues relating to China, technology and operations management, and corporate strategy. He has worked on executive education programs concerning strategic issues in China with a wide variety of corporations in the United States, Asia and Europe. He also contributes a regular column on business in China to www.cbiz.cn. Educated in engineering and business administration, Dr. Fischer became engaged in activities in China in 1980 as part of the academic team which established and ran the U.S. Commerce Department's Dalian program for senior Chinese officials and business leaders. He remained part of that program through 1984, and in 1999 was awarded a "visiting professorship" at the Dalian Institute of Technology. Since 1980, he has been involved in a series of activities that have kept him engaged in the Chinese development experience on a continuous basis for 25 years. From 1997 to 1999, he was both executive president and dean of the China Europe International Business School [CEIBS] in Shanghai, which is a joint venture between the European Commission and the Chinese government and probably the best-known business school in the country. Prior to this position and his present IMD appointment, he held the Dalton L. McMichael Sr. Professorship of Business Administration at the Kenan-Flagler Business School at the University of North Carolina at Chapel Hill where he was a faculty member from 1976 to 1996.

China remains a focal point of Dr. Fischer's research and teaching and he has recently been involved in several executive learning experiences in and about China, including programs for Deutsche Bank, Daimler-Chrysler, the Japan-American Institute for Management Science and the IMD EMBA program. He also continues to be involved in an active writing program regarding the Chinese economic reforms, Chinese industrial competitiveness and the foreign competitive and managerial experience in China. In 1999, Dr. Fischer received the Silver Magnolia award, Shanghai's highest honor for foreigners.

IMD, Chemin de Bellerive 23, 1001 Lausanne, Switzerland;
Phone: +41 21 618 0 328, Fax: +41 21 618 0 707, E-Mail: fischer@imd.ch

Dr. Yong Geng is an associate professor and associate director at the Institute for Eco-Planning and Development at Dalian University of Technology, P.R. China. He received a PhD in the field of industrial ecology at Dalhousie University, Canada and his special research interests include wastewater minimization, eco-industrial development, life cycle assessment tools and carrying capacity analysis. He was one of the initiators of a large project titled "Eco-Planning and Environmental Management in Chinese Coastal Communities", funded by the CIDA (Canadian International Development Agency) and implemented jointly by the University of Waterloo (Canada), Dalian University of Technology, Nanjing University and the Hainan Department of Land, Resource and Environment. This five-year project is dedicated to promoting China's environmental management capacity through demonstration projects, faculty and student exchanges, and joint research. Dr. Geng also led China's first eco-industrial park projects in Tianjin Economic Development Area (the largest industrial zone in China) and Dalian Economic and Technological Development Area (the second largest industrial zone in China). He has published a number of papers and books in the field of eco-industrial development, both in English and Chinese. In addition, he is a member of the International Society for Industrial Ecology, the Canadian Eco-Industrial Network and the Asian Eco-Industrial network.

School of Management Building, Dalian University of Technology, Dalian City, Liaoning Province 116024, P.R. China;
Phone: +86 411 8470 7331, Fax: +86 411 8470 8342,
E-Mail: ecoplan@dlut.edu.cn

Dr. Boris Gorella completed his PhD in chemistry at the Technical University of Berlin in 1991 and holds an MBA from INSEAD at Fontainebleau. He received the Feodor Lynen Award from the Alexander von Humboldt Foundation and was honored for an outstanding PhD thesis by the Fund of the German Chemical Industry. After completing his PhD he started working in the chemical industry in both Germany and the United States in the field of polymers. Prior to his current position he worked with McKinsey, where he was made Associate Principal. His assignments focused on strategic, organizational and operational projects as well as on post-merger integrations in various process and chemical industries in Germany and Asia. Dr. Gorella assumed his current position as head of Degussa Construction Chemicals, Business Unit Admixture Systems, Asia Pacific in 2002 and is based in Shanghai and Tokyo. Overall he has spent more than five years in different positions in Asia.

Degussa Construction Chemicals Asia Pacific, # 2301 Shanghai Times Square, 93 Huai Hai Zhong Lu, Shanghai 200021, P.R. China;
Phone: +86 21 5133 2830, Fax: +86 21 5306 9430,
E-Mail: boris.gorella@degussa.com

Dr. Klaus Griesar is global business development manager at Merck Electronic Chemicals. He started his career in the chemical industry in the Corporate R&D Strategy Department of SKW Trostberg in 1997. He joined Merck KGaA in 2000. Since then, he has headed various projects concerning divisional strategy and corporate business development. Since February 2004, as head of the Global Business Development Department of Merck Electronic Chemicals, he devoted most of his time creating new business opportunities for the division in Asia.

As vice president of the Association for Chemistry & Economics within the German Chemical Society Dr. Griesar has organized several conferences and workshops and has thereby contributed to establishing this association as a forum for discussion focusing on the economic aspects of the chemical industry in Germany. Dr. Griesar holds a teaching position at the University of Darmstadt, lecturing on companies, structures and strategies in the chemical industry.

Merck Electronic Chemicals Holding GmbH, Frankfurter Strasse 250,
64293 Darmstadt, Germany;
Phone: +49 6151 72 8134, Fax: +49 6151 72 91 8134,
Mobile: +49 151 1454 8134, E-Mail: klaus.griesar@merck.de

Dr. Christian Haug is co-founder of Bicoll GmbH, Munich, and was appointed the company's managing director in May 2002. He has also served as the Chairman of the Board of Directors of Bicoll Biotechnology (Shanghai) Co. Ltd. since July 2001. In addition, he is co-inventor of Bicoll's core technologies for discovering new lead structures. Prior to joining Bicoll, he did post-doctoral research work at the Shanghai Institute of Materia Medica, becoming the first German exchange researcher to receive the DAAD biosciences stipend in China. Before moving to the Far East, he completed a research placement in Central Research at Bayer AG, Leverkusen, where he worked in the solid-phase combinatorial chemistry group. Dr. Haug received his master degree and PhD from the University of Munich.

Dr. Haug, who was in charge of setting up Bicoll's facilities in Shanghai, is now responsible for Bicoll's research collaborations with international customers and technology development. His main research interest is the synthesis and derivatization of bio-active natural products and pharmacophores.

Bicoll Biotechnology (Shanghai) Co. Ltd., Bibo Road 518A,
Zhangjiang Hi-Tech Park, 201203 Shanghai, P.R. China;
Phone/Fax: +86 21 3895 3452, E-Mail: haug@bicoll-group.com

Dr. Gail E. Henderson is professor of social medicine, School of Medicine, and adjunct professor of sociology at the University of North Carolina at Chapel Hill, USA, where she has worked since 1983. Dr. Henderson holds a bachelor degree in Chinese language and literature from Oberlin College, and a master degree and PhD in sociology from the University of Michigan at Ann Arbor. She completed a post-doctoral fellowship in public health at the University of North Carolina. She has conducted research on health and health care in China since 1979, when she wrote her dissertation on a hospital in Wuhan (The Chinese Hospital: A Socialist Work Unit, with Myron S. Cohen, Yale University Press 1984). She is the author of numerous articles on health services in China. Most recently, she has worked with the China CDC on research ethics training for AIDS clinical investigators, and testified on China's rural health care system for the U.S. Congressional-Executive Committee on China (July 2003).

Department of Social Medicine, CB 7240, UNC School of Medicine,
Chapel Hill, NC 27599-7240, USA;
Phone: +1 919 843 8268, Fax: +1 919 966 7499,
E-Mail: ghenders@med.unc.edu

Dr. Christian Kober completed his PhD in inorganic chemistry at the University of Munich in 1993. He joined Degussa in 2004 and currently holds the position of director and head of Admixture Systems, China. Prior to this he was head of the Specialties Division of Wacker Chemie GmbH in Asia, where he set up a technical service center for Asia and established the basis for production in China. He also worked for two years in research evaluation in a key function at Wacker Chemie GmbH, as a trainee at Hoechst Japan Ltd. and as a Feodor Lynen scholar of the Alexander von Humboldt Foundation at the Tokyo Metropolitan University. He has spent more than 7 years in different positions in Japan and China. Dr. Kober has received numerous awards and stipends, including a Fulbright stipend to the USA, a DAAD ARC stipend to the U.K., a citation and an award for an outstanding PhD thesis from the Fund of the German Chemical Industry, a scholarship from the Japan Society for the Promotion of Science and a Carl Duisberg Society scholarship.

Admixture Systems North Asia, Degussa Construction Chemicals,
5/F Building 25, 69 Gui Qing Road, Caohejing Development Area,
Shanghai 200233, P.R. China;
Phone: +86 21 6485 3300 Ext. 822, Fax: +86 21 6495 0970,
Mobile: +86 137 01630310, E-Mail: christian.kober@degussa.com

Kuno Kohler is the regional president Asia North of Ciba Specialty Chemicals. He is based in Shanghai, the regional headquarters for Asia North, covering Mainland China, Hong Kong, Taiwan, Japan and Korea. He is a Swiss certified accountant and holds a master degree in Asian studies from the University of Leeds in the U.K. Having held a succession of finance positions within Ciba-Geigy in Australia, Taiwan and Indonesia from 1984 to 1994, Kuno Kohler returned to Taiwan to head the Group company in 1994. After the spin-off from Novartis, Kuno joined Ciba Specialty Chemicals and moved to Singapore to be the regional president of Asia South Region, overseeing the operations in South East Asia, Australia, New Zealand and South Asia before moving to China in 2004.

Ciba Specialty Chemicals (China) Ltd., 18th Floor Xin An Building,
99 Tian Zhou Road, Caohejing Hi-Tech Park, Shanghai 200233, P.R. China;
Phone: +86 21 2403 2101, Fax: +86 21 2403 2748,
E-Mail: kuno.kohler@cibasc.com

Dr. Andreas Kreimeyer is a member of the Board of Executive Directors of BASF Aktiengesellschaft. Andreas Kreimeyer was born in Hanover, Germany, in 1955. He studied biology at the Universities of Hanover and Hamburg. After being awarded his PhD, Kreimeyer started work in BASF's Main Laboratory in 1986, where his functions included the position of group leader for biotechnology. Starting in 1993, Kreimeyer was personal assistant to the Chairman of the Board of Executive Directors. In 1995, he transferred to Singapore, where he worked in the Southeast Asia/Australia division and his responsibilities included regional marketing, capital expenditures and strategy. In November 1998, Kreimeyer was appointed president of the Fertilizers division. He then became president of the Dispersions division in August 2000, which was renamed the Functional Polymers division in July 2001. Andreas Kreimeyer was appointed to the Board of Executive Directors of BASF Aktiengesellschaft with effect from January 1, 2003. Since February 1, 2003, his responsibilities comprise the Functional Polymers and Performance Chemicals operating divisions, and the Asia Pacific region.

BASF Aktiengesellschaft, E100, Room 1925, Carl-Bosch-Strasse 38,
67056 Ludwigshafen, Germany;
Phone: +49 621 60 92600, E-Mail: andreas.kreimeyer@basf-ag.de

Dr. Otto Kumberger completed his PhD in chemistry at the Technical University of Munich in 1992. He subsequently worked for one year as a post-doctoral fellow at the National Institute of Materials and Chemical Research in Tsukuba, Japan. In 1993 he joined the BASF ammonia laboratory, the company's chemical R&D competence center in Ludwigshafen. He worked there as a research chemist in heterogeneous catalysis R&D, as executive assistant to the head of the ammonia laboratory and as head of the controlling unit of the ammonia laboratory. In early 2001 he transferred to BASF East Asia Regional Headquarters in Hong Kong, where he took up an assignment as Senior Manager Asia Pacific for Investment & Strategies. In 2003 he became responsible for the regional business management of Diols & Polyalcohols in Asia Pacific. In January 2005 he started a new assignment as senior manager in the Strategic Planning department of BASF.

BASF Aktiengesellschaft, ZZS/A – D100, 67056 Ludwigshafen, Germany;
Phone: +49 621 60 40750, E-Mail: otto.kumberger@basf-ag.de

Dr. Kai Lamottke was appointed general manager and a member of the Board of Directors of Bicoll Biotechnology (Shanghai) Co. Ltd. in July 2001. He is co-founder of Bicoll GmbH, Munich and Managing Director of the company. In addition, Dr. Lamottke is co-inventor of Bicoll's technology for accelerating the drug discovery process. He also held a research position at the Technical University of Munich, sponsored by Bavaria's Ministry for Science, Research and Arts for two years. Prior to joining Bicoll he worked for a rational drug discovery company, 4SC AG, in constructing the combinatorial chemistry division. Before receiving his PhD from the University of Munich, he carried out the practical part of his diploma thesis at the University of California, Berkeley and received his master degree in chemistry from the University of Muenster. All his work was related to the synthesis and biosynthesis of medically relevant natural products.

Bicoll Biotechnology (Shanghai) Co. Ltd., Bibo Road 518A,
Zhangjiang Hi-Tech Park, 201203 Shanghai, P.R. China;
Phone/Fax: +86 21 3895 3452, E-Mail: lamottke@bicoll-group.com

Dr. Joachim E. A. Luithle studied chemistry at the University of Stuttgart, Germany and the Ecole Supérieure de Chimie Industrielle de Lyon, France. He earned his PhD from the University of Stuttgart in organic chemistry in the group of Prof. Jörg Pietruszka. He started his career at Bayer AG, working as a senior research scientist in medicinal chemistry at the company's Pharmaceutical Research Center in Wuppertal. Dr. Luithle then worked as a chemistry project leader. He is co-inventor of a development candidate for the treatment of Alzheimer's disease. He currently holds a position as Director of Innovation, HealthCare & Crop-Science in Corporate Development at the Corporate Center of Bayer AG and is assistant to the CTO.

Bayer AG, Corporate Center, Building W11, 51368 Leverkusen, Germany;
Phone: +49 214 30 36751, E-Mail: joachim.luithle.jl@bayer-ag.de

Dr. Heinz Mueller has been working at DZ Bank as an investment analyst for agrochemicals, consumer chemicals and chemicals stocks since April 1, 2000. He studied agriculture in Giessen, obtaining his master degree in agricultural engineering. He then worked on a research project commissioned by the German Research Association, which he completed with a PhD in agriculture. After successfully passing the examination as an agricultural assessor, he was employed from 1981 to 1995 in the Agro Division of Ciba-Geigy Deutschland (now Novartis/Syngenta). Here he worked from 1982 to 1991 as head of the strategic market research sector and EDP support. From 1991 on, he worked in Controlling. After that he moved to DG Bank where from 1995 to 1997 he was responsible for advising the Volks- and Raiffeisenbanken (local cooperative banks) in Schleswig-Holstein and Lower Saxony in agricultural matters. From 1998 to early 2000, he worked in Frankfurt in the Agriculture and Nutrition department.

DZ Bank AG Deutsche Zentral-Genossenschaftsbank, Equity Research,
Platz der Republik, 60265 Frankfurt am Main, Germany;
Phone: +49 69 7447 7947, Fax: +49 69 7447 2201,
E-Mail: heinz.mueller@dzbank.de

Marc P. Philipp is a business analyst in the Global Health & Life Sciences Practice of Accenture in Frankfurt, Germany. He is a member of Accenture's Strategy & Business Architecture Service Line where his functional focus covers corporate restructuring, market assessment and entry strategies, as well as mergers & acquisitions, with a strong affiliation to emerging markets. Before joining Accenture, he was engaged in the electronic and cord set supplier industry with regional focus on South-East Asia and China. He holds a master degree in business administration from the European Business School (Germany), the National University of Singapore and the Sid Craig School of Management at the California State University (Fresno, USA).

Accenture GmbH, Campus Kronberg 1, 61476 Kronberg, Germany;
Phone: +49 6173 94 66795, Fax: +49 6173 94 46795,
E-Mail: marc.p.philipp@accenture.com

Dr. Harald Pielartzik studied chemical engineering at the Technical University of the Lower Rhineland in Krefeld. After receiving his master degree in chemical engineering, he moved to the University of Dortmund to study chemistry. He received his PhD from the University of Saarland, Saarbruecken in 1984. The same year he joined Bayer's Central Research laboratories in Krefeld-Uerdingen, working on the development of liquid crystalline polymers. In 1989 he became group leader for new technologies in the Polycondensation Department of Central Research. From 1990 to 1994 he worked at Bayer's former U.S. subsidiary Mobay (now Bayer Corp.) in Pittsburgh, PA where he was responsible for the establishment of the Polymer Research Group. Upon his return to Germany in July 1994 he joined the Materials Research Department in Central Research as head of the Thermoplastics Group. In 1996 his research interest focused on functional materials and his department was re-named the Functional Materials Group. His R&D areas encompassed conductive polymers, materials for display technology as well as polymer materials for medical and diagnostic devices and drug delivery. From 2001 to June 2002 he was responsible for the Materials Design Department in Central Research. From July 2002 to December 2003 he was head of Bayer Polymers, Synthetic Rubber R&D Department with locations in Dormagen (Germany), Sarnia (Canada) and Orange (Texas). In 2004 he was named head of Bayer's Working Group on Nanotechnology with responsibility for all Bayer's development efforts in this field. Additionally he is responsible for coordinating Bayer MaterialScience's contacts to universities and associations.

Bayer MaterialScience AG, New Business Universities/Associations,
47812 Krefeld, Germany;
Phone: +49 2151 88 5790, Fax: +49 2151 88 7703,
E-Mail: Harald.Pielartzik@bayermaterialscience.com

Dr. Guido Reger is full professor for innovation and entrepreneurship at the University of Potsdam in Germany and director of the Brandenburg Institute for Entrepreneurship and Small and Medium-Sized Enterprises. Prior to this, he was professor at the University of Applied Sciences in Brandenburg and founding director of the Institute i.TEC – Innovation and Technology. From 1991 to 1998 he worked as a senior researcher at the Fraunhofer Institute for Systems and Innovation Research (ISI) in Karlsruhe. He was member of the Board of Directors of Fraunhofer ISI from 1996 until 1998. His research includes projects on industrial innovation strategies, globalization of research and technology, evaluation of science and technology policy, regional and national innovation systems. He has acted as a senior adviser to the German Ministry of Education, Science, Research and Technology (BMBF), the German Ministry of Economic Affairs (BMWi), the Swiss Federal Office for Education and Science (BBW), the OECD, the European Commission, and multinational corporations. From 1994 to 1998, he was coordinator and representative of the Federal Republic of Germany on the committee of the Innovation Program of the Commission of the European Communities. Guido Reger was visiting professor at the National Institute of Science and Technology Policy (NISTEP) in Tokyo, the Massachusetts Institute of Technology (MIT) in Cambridge, MA, the Beijing University of Aeronautics & Astronautics, and various European business schools. In April 2004, he received the research award of the International Association for Management of Technology (IAMOT) for his role as "... one of the most active and prolific researchers in the Technology Innovation Management field".

Guido Reger studied sociology, economics and management at the Hamburg University of Economics and Politics. He obtained a PhD from the University of St. Gallen on coordination and strategic management of international innovation processes.

Department of Economics and Social Sciences, University of Potsdam,
August-Bebel-Strasse 89, 14482 Potsdam, Germany;
Phone: +49 331 977 3340 (3317), Fax: +49 331 977 3619,
E-Mail: reger@rz.uni-potsdam.de

Stefan Sommer is currently president of DSM (China) Ltd. DSM is a global leader in specialty chemicals and active in the fields of life science products, nutritional products, performance materials and industrial chemicals. DSM generated revenues of around EUR 8 billion in the year 2003 (pro-forma, incl. DSM Nutritional Products) and employs about 25,000 people worldwide. In China, DSM operates 11 wholly owned entities and joint ventures, employs more than 3,000 people and generates revenues of more than EUR 400 million. Before moving to DSM, Stefan Sommer spent more than 7 years with Hoechst AG in Frankfurt am Main, Germany; first in Corporate Strategy and then in Mergers & Acquisitions. After the demerger of Celanese AG from Hoechst, he joined Ticona, the engineering plastics business of Celanese AG, and became the company's President in 2002. Ticona generated revenues of some EUR 750 million in 2002 and employed around 2,400 people worldwide. During his time with Ticona he also served as director on the boards of three Ticona joint ventures in Asia.

From 1991 to 1996 Stefan Sommer worked as a consultant, project manager and associate director at Arthur D. Little Management Consultants in Munich,

Germany. From 1978 to 1982 he completed a commercial apprenticeship with Kloeckner Industrie-Anlagen GmbH in Duisburg, Germany. He then worked as market segment manager for Kloeckner in Jakarta, Indonesia, before studying business administration in Munich. He started his professional career with Hoechst AG in Frankfurt as a product manager for specialty chemicals.

DSM (China) Ltd., 22F Ocean Towers, 550 Yan An Road (East), Shanghai 200001, P.R. China;
Phone: +86 21 3310 4988, E-Mail: stefan.sommer@dsm.com

Dr. Elmar Stachels is managing director of Bayer China Company Ltd. and has been Greater China Country Group Speaker for Bayer since January 1, 2001. He previously headed Bayer's Asian Regionalization Project in 2000, which directly resulted in the creation of the Bayer Group for Greater China, consisting of China, Hong Kong and Taiwan.

Dr. Stachels was born on in Coesfeld, Germany. He studied jurisprudence and economics in Tuebingen, received his PhD and had an assistant position at the University of Cologne. His PhD thesis (judged summa cum laude) about the law of stability and growth ("Das Stabilitaets- und Wachstumsgesetz im System des Regierungshandelns") was published by Walter de Gruyter Verlag. His career with Bayer started in 1973 in the company's Legal Department in Leverkusen. In 1979 he took over responsibility as managing director and national spokesman for Bayer in Iran until 1982, when he returned to Germany to become head of the sales department of the Specialty Polyurethane Division. He went on to take up the position as managing director and national spokesman of Bayer in Mexico from 1988 to 1992. In 1993 he returned to Germany to become head of marketing of Bayer's dyestuffs business. He held this position until 1995, when he became a managing director of DyStar, Bayer's joint venture with Hoechst in Frankfurt, Germany.

Bayer (China) Ltd., Corporate Communications, 34F Jing Guang Centre, Beijing 100020, P.R. China;
Phone: +86 10 6597 3181 Ext. 2801, Fax: +86 10 659 73292,
E-Mail: elmar.stachels.es@bayer-ag.de

Dr. Martin S. Vollmer is currently head of Aromatic Coatings, Resins & Specialties in the Innovation Department of the Coatings, Adhesives, Sealants Business Unit at Bayer MaterialScience AG. Prior to this, he worked in Corporate Development at Bayer AG as assistant to the CTO and was involved in several new business development and technology sourcing initiatives, with a special focus on China. He started his industrial career in Bayer's former Central Research Division as head of a materials research laboratory.

Martin Vollmer studied chemistry at the University of Stuttgart and obtained his PhD at the Institute for Organic Chemistry working with Prof. Franz Effenberger in the field of molecular electronics. After completing his PhD he did post-doctoral work at the Scripps Research Institute in La Jolla, California where he was involved in biomaterials and nanotechnology research programs.

Bayer MaterialScience AG, Building Q1, 51368 Leverkusen, Germany;
Phone: +49 214 30 27958, Fax: +49 214 30 50708,
Mobile: +49 175 30 27958,
E-Mail: martin.vollmer@bayermaterialscience.com

Angela Wang is the head of Strategy and Business Development at Beijing Novartis Pharma, Ltd. in China. Prior to joining Novartis AG in 1996 as Country Head of Finance for Novartis Group in China, she had been with Arthur Andersen and Deloitte Touche in the United States. She holds a bachelor degree in financial accounting from Rutgers University in New Jersey, a master degree in business administration from IMD in Lausanne, Switzerland and a CPA qualification in New Jersey and California. After several years in Novartis headquarters in Basel, Switzerland and in China, her professional background includes functional and operational experience in the development of new businesses as well as the optimization of existing operations for Novartis Group. In China, her areas of expertise include establishing business entry, the setting up of new joint ventures and trading companies, managed restructuring and turnaround of businesses, negotiations with joint venture partners on the closure of unprofitable ventures, and managed wind-down. Currently, she focuses on developing long-term strategy for growing Novartis' pharmaceutical business in China and generating new business opportunities in this fast-growing market.

Beijing Novartis Pharma Ltd., China World Tower 2, 1 Jianguomenwai Ave,
Beijing 100004, P.R. China;
Phone: +86 10 6505 8833, Fax: +86 10 6505 6992,
E-Mail: angela.wang@pharma.novartis.com

Dr. David E. Webber is a consultant to the pharmaceutical industry and acts as the director general of the World Self-Medication Industry (WSMI). WSMI is a federation of over 50 member associations representing manufacturers and distributors of non-prescription medicines on all continents. WSMI offices are located in France, near to Geneva. The dual focus of WSMI is to support the development of responsible self-medication worldwide through the network of member associations, and to work with multilateral organizations in areas of common interest. Dr. Webber has been particularly involved in helping develop the pharmaceutical industry in emerging countries in Latin America, in the Middle East and in China.

Dr. Webber was IFPMA fellow and director of economic policy from November 2000 to November 2002, on secondment from Glaxo SmithKline plc to the International Federation of Pharmaceutical Manufacturers Associations (IFPMA) in Geneva, Switzerland. During this time a basic R&D model ("Encouraging Pharmaceutical R&D in Developing Countries") and a China case study ("Accelerating Innovative Pharmaceutical Research and Development in China") were developed. Support for the China case study and for the book chapter was provided by RDPAC, the R&D-Based Pharmaceutical Association in China. Dr. Webber's work in China has continued in his current role, most recently through co-organizing the WSMI 6th Asia Pacific Regional Conference in Beijing in October 2004 under the theme "From the Past to the Future: Public Health, Regulatory and

Industrial Opportunities in Responsible Self-Medication". In representing the pharmaceutical industry at the global level, Dr. Webber has worked on various industry and joint industry-multilateral working groups, with the WHO, the World Bank and the Commission for Macroeconomics and Health. Dr. Webber has published various papers on pharmaceutical R&D, manufacturing and marketing topics and in 2000 wrote a book for Sir Richard Sykes, Chairman of Glaxo Smith-Kline, entitled 'New Medicines, The Practice of Medicine and Public Policy, published by the Nuffield Trust (ISBN 0-11-702676-X).

Dr. Webber has a PhD in animal physiology from London University and undertook post-doctoral research in gastric physiology. He joined the pharmaceutical industry in 1982 and held a series of increasingly senior positions, latterly as a Director of Corporate Strategy for Glaxo Wellcome with particular responsibility for mergers and acquisitions, competitive analysis and corporate strategy.

WSMI, C.I.B. – Immeuble A – "Keynes", 13 chemin du Levant,
01210 Ferney-Voltaire, France;
Phone: +33 45 02 84 725, Fax: +33 45 02 84 024, E-Mail: dwebber@wsmi.org

Jörg Wuttke was born in Ulm, Germany. On completion of his bachelor degree in business administration and economics in Mannheim, Germany, he studied Chinese at a language institute in Taiwan in 1984. He joined the finance department of BBC (from 1988 ABB) in Mannheim in 1985 and worked as the commercial manager on a construction site in Jakarta, Indonesia from the end of 1986 to 1987. He was transferred to Beijing in 1988 as finance and administration manager for ABB Beijing. In 1990 he returned to Germany as sales manager of ABB Power Plants Division, Mannheim, responsible for Africa, Russia and Germany. In 1993 he returned to China as chief representative ABB China, Shanghai Office. In order to capture several large projects he was transferred in 1994 to the President's Office of ABB China, Beijing and took on responsibility for the development and financing of large projects. In this period of time two large power projects in Ulaan Baator and Beijing were finalized. In early 1997 he was appointed general manager and chief representative of BASF China in Beijing.

In addition to his job he is active in several associations. In 2000 he was one of the founding members of the European Chamber of Commerce in China and became the first chairman of its petrochemical working group. In 1999 the German Chamber of Commerce in China was founded in Beijing and he joined its first board. In early 2001 he was elected by the three boards in Beijing, Shanghai and Guangzhou as their Chairman and served until April 2004. The Chamber has more than 800 member companies across China. Mr. Wuttke is a very active member of the Rotarians in Beijing, raising money and supervising five educational projects in Qinghai and Tibet province. He is also a member of European Bahai Business Foundation since 1993.

BASF China Ltd., Beijing Sunflower Tower 15th Floor,
No. 37 Maizidian Street Chaoyang District, Beijing 100026, P.R. China;
Phone: +86 10 6591 8899 Ext 6001, Fax: +86 10 85275599,
E-Mail: wuttkej@basf-china.com.cn

Dr. Dahai Yu completed his PhD in organic chemistry at the University of Hamburg in 1989 and did his post-doctoral research at the Centre National de la Recherche Scientifique in France.

He is now vice president and general manager of Degussa's agrochemicals and intermediates business. The general focus of his work is to turn around and reshape the predominantly Europe-based business and develop the respective business in Asia, particularly in China. Prior to this assignment, he held various management positions in R&D, sales and marketing for new products, technology transfer and corporate development within Degussa. His last assignment was Senior Vice President of Controlling of the Fine and Industrial Chemicals Division. He has spent about half of his life in Europe and is engaged in many cross-cultural activities.

Degussa AG, Dr.-Albert-Frank-Strasse 32, 83308 Trostberg, Germany;
Phone/Fax: +49 86 1286 3011/13, Mobile: +49 171 1661405,
E-Mail: dahai.yu@degussa.com

Dr. Maximilian von Zedtwitz is a professor of technology and innovation management at Tsinghua University, Beijing, P.R. China, and director of the Research Center for Global R&D Management with locations in Beijing and St. Gallen, Switzerland. He teaches in MBA, PhD and executive education programs. He worked for Siemens, ATR-NTT, the University of St. Gallen and Harvard University before joining IMD-International (Lausanne, Switzerland) in 2000 as a professor of technology management. He holds master and bachelor degrees in computer science, and MBA and PhD degrees in business administration.

Maximilian von Zedtwitz's research focuses on R&D management, in particular transnational R&D organization and technology-based entrepreneurship. He has published five books and more than forty articles in leading practitioner and academic journals in English, German, French, Japanese and Chinese. He has also been cited in *The New York Times,* the *International Herald Tribune, Der Spiegel, Le Temps* and *NZZ.* He won awards for best papers from the *R&D Management Journal* in 1998, *ISMOT* in 1998 and *ABB* in 2001. He currently serves on the editorial boards of *R&D Management, Journal of International Management, Technological Analysis and Strategic Management* and three other international journals. Maximilian von Zedtwitz is an executive director at *AsiaCompete*, a training and consulting company based in Hong Kong, and an executive director of *IAMOT*, the International Association for Management of Technology. He is also on the advisory boards of *GetAbstract,* an Internet-based book summary provider, *Squarewise*, a Dutch consulting company, and *INSEAD InnovAsia*, an information services company in Singapore. In consulting, he concentrates on entrepreneurship, R&D and innovation strategy, and start-up incubation.

Mailbox B-55, School of Economics & Management, Tsinghua University,
Beijing 100084, P.R. China;
Phone: +86 10 6278 9797, Fax: +86 10 6278 2253,
Mobile: +86 136 1103 1522, E-Mail: max@post.harvard.edu

Literature

Abelshauser W, von Hippel W, Johnson JA, Stokes RG (2004) German Industry and Global Enterprise, BASF: The History of a Company, Cambridge University Press, Cambridge

Arora A, Landau R, Rosenberg N (1998) Chemicals and Long Term Economic Growth. Wiley

Arnold HM, Lamottke K (2003) Erfolg im Abwärtszyklus – Geschäftsmodelle, die vom schwierigen wirtschaftlichen Umfeld profitieren können. Sonderausgabe Biotechnologie, GoingPublic Verlag, pp 122-123

Asia Pulse Businesswire (2004) Electronic Material Industry and Market in China. August 11, 2004

BASF Aktiengesellschaft (1994) BASF: Milestones in its History

BASF and Chinese Academy of Engineering (2002) Proceedings on Symposium on Energy, Environment, Sustainable Development

Becker J (2000) The Chinese. John Murray, London

Beckmann (2004) China: Die aufstrebende Wirtschaftsmacht. Deutsche Bank Global Markets Research, Frankfurt

Business Daily Update (2004a) China's Computer Chip Market Hits US$24,99b in 2003. February 23, 2004

Business Daily Update (2004b) Semiconductor Industry to Get Boost in China. March 22, 2004: p 22

Chang-Xiao Liu, Pei-Gen Xiao (2002) Recalling the research and development of new drugs originating from Chinese traditional and herbal drugs. Asian Journal of Drug Metabolism and Pharmacokinetics 2: 133-156

Chemical Engineering News (2002) Electronic Chemicals. July 15, 2002: p 19

Chemical Week (2003) Electronic Chemicals. June 23, 2003

Chen S (2003) Comparative Study on Pharmaceutical Industries between China and India. In: China Pharmaceutical Daily, July 2 (in Chinese)

Chen W (2004) Looking For an Industry Cure. China International Business, Issue 204, November 2004, pp 34-37

China Daily (2003) Internet edition, November 12, 2003

China Economic Information Network (2003) China's Pharmaceutical Industry 2003. Beijing

China Post (2004) China 300-mm wafer plant to open. September 15, 2004

Chuan Q (2004) GlaxoSmithKline confident of Chinese market. The China Daily November 23rd

Cochran S (2000) Encountering Chinese Networks: Western, Japanese, and Chinese Corporations in China, 1880-1937. University of California Press, Berkeley

Cohen MS (1984) The Great Cephalosporin Wars. European Journal of Clinical Microbiology 3: 177-179

Cooke M (2003) Diverse views on the future. European Semiconductor, April 2003: 25-30

Davenport RE, Fink U, Yoshio I (2004) Electronic Chemicals. SRI Consulting

Dow Jones News Services (2004) Chinas Chipmaker Market Will Be $100 Billion by 2008. February 16, 2004

Downton C, Clark I (2003) Statins – the heart of the matter. Nature Reviews Drug Discovery 2, pp 343-344

Easton R (2003) Supply Chain Management in China: Assessments and Direction. Ascet Volume 5

Electronic Engineering Times (2003) October 27, 2003

Electronic News (2004a) IC demand on the Rise in China. March 29, 2004

Electronic News (2004b) China denies overcapacity, unfair trade allegations. March 22, 2004

Enriquez J, Goldberg R (2000) Transforming Life, Transforming Business: The Lifescience Revolution. Harvard Business Review, March-April, 96-104

Espicom Business Intelligence (2004) China Market Study. West Sussex

Feng C (2002) Risks and Rewards for Pharma in Post-WTO China. Pharmaceutical Executive April 1st 2002

Festel G (2003a) Historical Evolution and Actual Trends of the Global Chemical Industry. Chemie & Wirtschaft, Jg. 2, Ausgabe 1, Februar 2003, 27-34

Festel G (2003b) Shareholder Value as Driving Force for Transformations of Chemical Companies. Chemie & Wirtschaft, Jg. 2, Ausgabe 4, November 2003, 26-42

Festel G, Knöll J, Götz H, Zinke H (2004) Bio-ways to drugs. European Chemical News. 19-25 January 2004, 22

Financial Times UK (2004a) Speedy Rise of China's Top Chip. November 2, 2004: p 29

Financial Times UK (2004b) Fear of High-tech Piracy Makes some Microchip Companies Cool about China. July 15, 2004: p 11

Forney M (2003) China's Failing Health System. Time Asia, Beijing

Fukuyama F (1995) Trust. Hamish Hamilton

Frankfurter Allgemeine Zeitung (2004) Taiwanesische Unternehmen locken Infineon. March 23, 2004: p 24

Geng Y, Côté R (2001) EMS as an Opportunity for Enhancing China's Economic Development Zones: the Case of Dalian. In: Environmental Management for Industrial Estates-A Resource & Training Kit. UNEP Division of Technology, Industry and Economics, Production and Consumption Branch, Paris

Geng Y, Côté R (2003) Environmental Management System at the Industrial Park Level. Environmental Management, Vol. 31, No. 6, 784-794

Global Sources (2004) Consumer Electronic Outlook (see also: www.globalsources.com)

Guo D (2003) How Overseas Companies Invest into China's Pharmaceutical Market. www.healthoo.com (in Chinese)

Haug C, Lamottke K (2004) Pionierarbeit – auch in der Mitarbeiterführung. ChinaContact, 1+2, pp 29-40

Hein C (2004) Warnschuss aus China. Frankfurter Allgemeine Zeitung 290: p 11

Henderson GE et al. (1998) Trends in Health Services Utilization in Eight Provinces in China, 1989-1993. Social Science and Medicine. 47(12): 1957-1971

IMS Health (2004) Market prognosis China 2004-2008. London

Jackson R, Howe N (2004) The graying of the middle kingdom. Center for Strategic and International Studies, Research publication, April 2004

Joseph G and Schaefer R (2004) China: Auf dem Weg zur Wirtschaftsgroßmacht. Economic Research Allianz Group Dresdner Bank Working Paper, Frankfurt

Jungmittag A, Reger G, Reiss T (eds.) (2000) Changing Innovation in the Pharmaceutical Industry – Globalization and New Ways of Drug Development. Berlin, New York et al.

Lampton DM (2003) China's Health Care Disaster. Asia Wall Street Journal May 6th, 2003

Langlois JD (2001) The WTO and China's Financial System. China Quarterly, Vol. 167, 610-629

Liao J (2003) The Development Route of Western Pharmaceutical Enterprises. www.mediaChina.net (in Chinese)

Liu S and MacKellar L (2001) Key Issues of Aging and Social Security in China. International Institute for Applied Systems Analysis, Laxenburg

Liu Y, Rao K and Hsiao WC (2003) Medical Expenditure and Rural Impoverishment in China. Centre for Health and Population Research, Harvard University, Boston

MacKinnon R (2004) Kaliumkanäle und die atomare Basis der selektiven Ionenleitung (Nobel-Vortrag). Angew. Chem. 116, 4363-4376

Mandell GL, Douglas RG, Bennett JE (1990) Principles and Practice of Infectious Diseases. Churchill Livingstone, New York

Moore G (1997) An Update on Moore's Law. Intel Developer Forum, Keynote. San Francisco. Sep. 30, 1997

Newman DJ, Cragg GM, Snader KM (2003) Natural Products as Sources of New Drugs over the Period 1981-2002. J. Nat. Prod. 66, 1022-1037

News Report (2003) How to handle upcoming wave of large volume of pharmaceutical patents to be terminated. In: China Industrial Commerce Daily, February 2003 (in Chinese)

Porter ME (1980) Competitive Strategy. Free Press, New York

Prahalad CK and Lieberthal K (1998) The End of Corporate Imperialism. Harvard Business Review, Boston

RDSL Asia (2004) China – Chip Industry Sees Growth in Production, Investment. February 17, 2004

Reddings G (1993) The Spirit of Chinese Capitalism. De Gruyter

Reed Electronics (2004) www.reed-electronics.com/semiconductor/contents/pdf/SIBreakfast2004.pdf

San Jose Mercury News (2004a) China, Partner to U.S. Tech Firms, is also a Fast-growing Thread. March 14, 2004: p 1

San Jose Mercury News (2004b) U.S. High-Tech Giants Invest in Future Competitor. March 15, 2004

Schmitt S (2002) Der Arzneimittelmarkt der VR China. German-Chinese Business Forum 02/2002: 7-12

Schmitt S (2002a) Arzneimittelvertrieb – VR China lässt ausländische Unternehmen zu. German-Chinese Business Forum 02/2002: 28-30

Sharper I (2002) All change for Chinese pharma. Scrip Magazine 12/02: 6-7

South China Morning Post (2004a) SMIC Sales to Grow on Increased Capacity. February 14, 2004: p 4

South China Morning Post (2004b) Chipmakers Risk Capacity Oversupply. March 24, 2004: p 2

Special Report (2004a) Strategies of Multinational Pharmaceutical Companies in China. www.hc360.cn (in Chinese)

Special Report (2004b) Development Analysis on China's FDI Firms in Pharmaceutical Industry behind the Trend of Globalization. www.jinhf.com (in Chinese)

Taiwan Economic News (2004) SMIC to unveil mainland China's first 300mm wafer fab soon. September 20, 2004

The Guardian (2003) Dismal fate awaits China's millionaires. October 31, 2003

Time Magazine (2004) August 9, 2004: p 13

Torreblanca M (2004) The China syndrome. Scrip Magazine 04/04: 28-29

Trinh T (2004) Deutsche Investitionen in China – Chance für deutsche Unternehmen? Deutsche Bank Research, Frankfurt

United Nations Environment Programme (UNEP) (2001) Environmental Management in Industrial Estates in China. Paris, France: United Nations Environment Programme, Division of Technology, Industry and Economics. p 95

Vaishampayan P and Chen VY (2004) Will India and China dominate pharmaceutical manufacturing? UBS Investment Research – Q-Series: Asian Pharma Sector

Venture Capital Journal (2004) Road to Riches? June 1, 2004

Vernon R (1966) International Investments and International Trade in the Product Life Cycle. In: Quarterly Journal of Economics, May 1966, pp 190-207

Vernon R, Wells LT (1986) The Economic Environment of International Business, the 4th edition, Englewood Cliffs, NJ; Prentice Hall

Von der Hagen W, Gruss A and Wolff B (2002) Das Wachstumspotenzial im Gesundheitswesen in China. German-Chinese Business Forum 02/2002: 23-27

Wang X (2002) Impact on the Chinese Pharmaceutical Industry after Entering into the WTO. In: China Chemical Engineering Information, July 2002 (in Chinese)

Watanabe M (2002) Holding company risk in China: a final step of state-owned enterprises reform and an emerging problem of corporate governance. China Economic Review, December 2002, Vol. 13, No. 4, 373-381(9)

Webber DE (2003a) Accelerating Innovative Pharmaceutical Research and Development in China: A Case Study. International Federation of Pharmaceutical Manufacturers Associations, Geneva Switzerland

Webber DE (2003b) Encouraging Pharmaceutical R&D in Developing Countries. International Federation of Pharmaceutical Manufacturers Associations, Geneva Switzerland

White RS (1998) Technological Systems and Organizational Change: China's Pharmaceutical Industry in Transition. Ph.D. Dissertation, Sloan School of Management, Massachusetts Institute of Technology

WHO (The World Health Organisation) (2002) Traditional Medicine – Growing Needs and Potential. WHO Policy Perspectives on Medicines. No. 2 May 2002. WHO, Geneva Switzerland

World Bank (1997) China 2020: Financing Health Care: Issues and Options for China

Yang JT (2001). The Environmental Management of Industrial Estates in China: Issues and Perspectives. Paper presented at the Environmental Management of Industrial Estates Workshop, 8-11 July, Dalian, China

Yang X (2004) Developing Fine Chemical Industry within Chemical Parks in China. China Petroleum and Chemical Industry, No. 3, 62-65

Zheng X (2004) Implementation of Classified Drug Administration System Gives Full Play to Self-medication in Health Care. Proceedings of the World Self-Medication Industry 6th Asia Pacific Conference, Beijing China, October 2004. In press

Index